VII International Conference: Modern Technologies for Non-Destructive Testing 2018

IOP Conference Series: Materials Science
and Engineering Volume 457

Tomsk, Russia
8 – 13 October 2018

ISBN: 978-1-5108-8003-0
ISSN: 1757-8981

Printed from e-media with permission by:

Curran Associates, Inc.
57 Morehouse Lane
Red Hook, NY 12571

Some format issues inherent in the e-media version may also appear in this print version.

Printed by Curran Associates, Inc. (2019)

For permission requests, please contact the Institute of Physics at the address below.

Institute of Physics
Dirac House, Temple Back
Bristol BS1 6BE UK

Phone: 44 1 17 929 7481
Fax: 44 1 17 920 0979

techtracking@iop.org

Additional copies of this publication are available from:

Curran Associates, Inc.
57 Morehouse Lane
Red Hook, NY 12571 USA
Phone: 845-758-0400
Fax: 845-758-2633
Email: curran@proceedings.com
Web: www.proceedings.com

TABLE OF CONTENTS

IOP Conf. Series: Materials Science and Engineering **457** (2018) 011001 doi:10.1088/1757-899X/457/1/011001

1. Abstract

This issue of the journal is dedicated to the scientific and applied research presented at the VII International Conference "Modern Technologies for Non-Destructive Testing". The conference was held on October 8–13, 2018 in Tomsk (Russia).

The conference brought together experts from different countries and organizations who received a great opportunity to share knowledge in sessions, to initiate discussions on topics of interest to all participants.

The aim of the Conference was to support young talented people, to involve pupils, students, post-graduate students and young scientists in the solution of actual scientific problems of non-destructive testing and technosphere safety.

The main goals of the event were exchange of information and discussion of topics related to non-destructive testing, technosphere safety and the environment and widespread implementation of the obtained results.

The key topics of the conference were:

– Non-Destructive Testing Methods (Sensor Concepts, Algorithmics, Methods Combination, New Methods, Standards and Training);
– Modelling and Data Processing (NDT Modeling & Computer Visualization, Simulations);
– Technological Processes and Systems (Sensors; Data Processing; Modeling of Technological Processes);
– Materials Science (Metals and Alloys; Engineering; Semi-finished Products: Pipes, Plates, Bars, Composites; Fundamentals, Characterization and Computational Modeling);
– Technosphere safety (testing in the field of environmental protection, environmental management monitoring, oversight functions in the field of recycling and use of industrial waste, ensuring human safety in everyday life and emergency situations.

The Conference Organizer is Division for Testing and Diagnostics of the School of Non-Destructive Testing & Security of the National Research Tomsk Polytechnic University.

2. Committee:

Aleksandr S. Sigov - Dr., Prof., academician, president of Moscow State Institute of Radio-Engineering Electronics and Automation , general chair, Moscow, Russia
Eduard S. Gorkunov - Dr., Prof., academician, director of Institute of Engineering Science, UB RAN, Yekaterinburg, Russia
Anatoliy P. Surzikov - Dr., Prof., Head of Department of Division Tomsk Polytechnic University, vice-chairman, Tomsk Russia
Boris I. Kapranov - Dr., Prof., Sector leader Technical Tomography and Introscopy Tomsk Polytechnic university, vice-chairman, Tomsk, Russia
Eduard Siemens - Dr.-Ing., Prof., Anhalt University of Applied Sciences, Köthen, Germany.
V.A. Pomogaev - researcher of Radiation Lab, postdoc of Chemical and Biomedical engineering department, Notherdam University, Indiana, USA
Jao Ven - Prof., Northeastern University, Massachusetts, USA
Thomas Bohlen - Dr., Prof., Karlsruhe Institute of Technology, Karlsruhe, Germany
André Kurzmann - Institute of geophysics, Karlsruhe Institute of Technology, Karlsruhe, German
Klaus Eichhorn - researcher, Institute of learning photons and synchrotron emission, Karlsruhe Institute of Technology, Karlsruhe, Germany
Oldrih Stary - Faculty of Electrical Engineering, Czech Technical University in Prague, Prague, Czech Republic
S.B. Habibova - Chief of International Relations Department, The Ministry of Education of the Republic of Tajikistan, Tajikistan
Ilkay Yilmaz - Prof., vice-chairman, Department of Economics, Mersin University, Turke
B.T. Torobekov - PhD, professor, vice-rector for academic work Kyrgyz State Technical University after I. Razzakov , Kyrgyzstan
Aleksandr V. Rogachev - PDr., Prof., Rector of Francisk Skorina Gomel State University, Gomel, Republic of Belarus
Alexey Elyashevich - Dr., Prof., Peter the Great St.Petersburg Polytechnic University, St.Petersburg, Russia
Petr Ya. Kalita - DPhD, academician of Ukraine academy, president of Ukraine quality association, Academy of Quality Problems, Ukraine
Tatiana V. Fursa - Dr., leading researcher Electronics, Dielectrics and Semiconductors Lab Tomsk Polytechnic university, Tomsk, Russia
Inna V. Plotnikova - PhD, Associate Professor of Department of Division for Testing and Diagnostics Tomsk Polytechnic university, Tomsk, Russia
Irina A. Larioshina - PhD, Associate Professor of Department of Division for Testing and Diagnostics Tomsk Polytechnic university, Tomsk, Russia

IOP Conf. Series: Materials Science and Engineering **457** (2018) 011001 doi:10.1088/1757-899X/457/1/011001

Alexey N. Kalinichenko - PhD, Associate Professor of Division for Testing and Diagnostics Tomsk Polytechnic university, conference chair, Tomsk, Russia

3. The Organizing Committee of the VII International Conference "Modern Technologies for Non-Destructive Testing" wishes to thank the IOP Conference Series Materials Science and Engineering-Team for collaboration.

IOP Conf. Series: Materials Science and Engineering **457** (2018) 011002 doi:10.1088/1757-899X/457/1/011002

Peer review statement

All papers published in this volume of *IOP Conference Series: Materials Science and Engineering* have been peer reviewed through processes administered by the proceedings Editors. Reviews were conducted by expert referees to the professional and scientific standards expected of a proceedings journal published by IOP Publishing.

Analysis of washing modes of metal tin cans outer surface in jet-type washing machines

A V Mayorov[1], N V Yanukov[1], D A Mikheeva[2], N E Yaytseva[3], D V Lukina[1]

[1]Associate Professor, Mari State University, Yoshkar-Ola, Russia
[2]Graduate Student, Mari State University, Yoshkar-Ola, Russia
[3]Student, Mari State University, Yoshkar-Ola, Russia
E-mail: ao_maiorov@mail.ru

Abstract. The main parameters of jet washing of cans are: the concentration of the alkaline solution and its temperature, the pressure of the solution at the nozzle, the shape and dimensions of the nozzle opening, the angle of the direction of the jet toward the surface being cleaned, the distance from the water jet nozzle to the surface of the washable cans. The study of these parameters was carried out on a special experimental setup equipped with instrumentation. Dependence of washing efficiency on solution pressure given to the cap with different diameters of its holes was found. Furthermore, it is established that the process efficiency decreases with increase of pressure for more than 3 kg/cm^2, as a certain hydrodynamic mode of jet percussion action begins at the boundary with the surface being cleaned. Cleaned surface becomes larger with increase of concentration and temperature of washing solution. Increase of solution concentration for more than 3% and its temperature for more than 70÷75°C do not promote intensification of the process. Cleaned surface depends on angle of jet direction at the point of its strike with the can surface. This dependence represents the fact that the cap should be placed in relation to transported cans in such a way that the jet would be always at right angle to the dirty surface.

1. Introduction

Canned goods are one of the most consumed products in the Russian Federation. There are several reasons of this phenomenon: wide range of canned food, gain of consumers' welfare, increase in demand on ready meals due to acceleration of life pace, extended storage period of canned goods, special storage conditions are not required, ease of use, contraction of home-made canned food [1–3].

Every year, in the Russian market of tin cans appear more and more new trademarks, new manufacturers, which aspire to improve the quality of produced product and attract a buyer. In this situation, the packaging and label of the goods are powerful means of promoting the goods on the market.

Package and label are important for promotion of the product in the market. That is why it is necessary to stick the clean undamaged label on the whole surface of the tin can firmly and accurately.

According to the technical requirements, an automatic labeling machine should work with the clean dry tin cans having no grease stains on the surface and meeting the requirements of the GOST 5981-88 and GOST 11771-93. The procedure of cans degreasing with the help of jet-type washing machines was included in the technological process of canned goods production to consider all these requirements.

Quality of cans cleaning, labour capacity, industrial standards, and cost price of canned food production depend on excellence of technology and washing machines.

A large number of modern washing machines work in an irrational regime and are energy and metal consuming, so their use is unprofitable. Consequently, the development of new efficient machines and the improvement of existing washing machines in the canning industry is great reserve to reduce energy consumption, materials and cost of the whole production process of canned food.

Therefore, research aimed at improving the washing process filled cans in the production of canned food are very relevant and have great importance in the food industry.

Appliance of jet and submersible cleaning as the less energy-demanding method is more prospective according to the criteria of specific energy costs on a unit of the surface being cleaned [4–7].

2. Experimental

The main parameters of jet washing of cans are: the concentration of the alkaline solution and its temperature, the pressure of the solution at the nozzle, the shape and dimensions of the nozzle opening, the angle of the direction of the jet toward the surface being cleaned, the distance from the water jet nozzle to the surface of the washable cans. The study of these parameters was carried out on a special experimental setup equipped with instrumentation.

The time of action of the detergent solution on the cleaned surface was measured with a stopwatch.

When studying the basic parameters characterizing the washing process of cans, a method of assessing the quality of washing was applied, which consists in exposing the surface of the cans to ultraviolet irradiation. At the same time, the remnants of non-washed contaminants (mainly fats) start to lumi- nume, i.e., to glow. It is easy to fix it not only visually, but also by photographing.

3. Results of Researches

The results of experimental research specifying the main parameters and washing modes of the filled metal tin cans outer surface with the help of the jet-type washing machines are represented below.

Experiments allowed to determine efficiency of washing, that is dependence of the cleaned surface on solution concentration and time of exposure on this surface. From the diagram (Figure 1), it can be seen that the cleaned surface becomes larger with increase of solution concentration. The process starts intensively and then it slows down. Optimal solution concentration is 2÷3%.

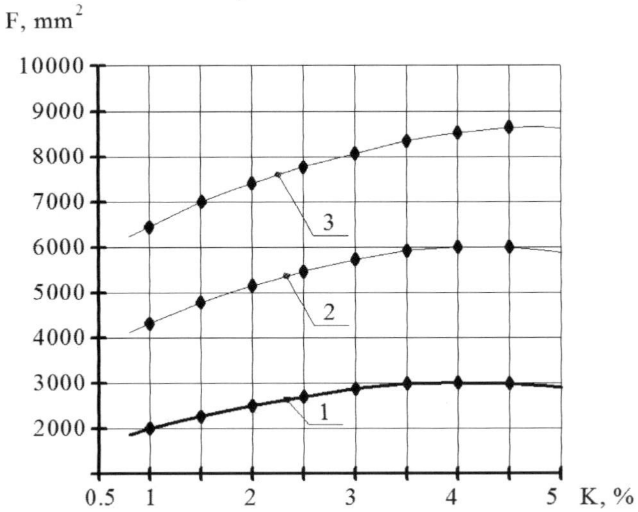

Figure 1. Dependence of the cleaned surface F on concentration of solution K of caustic soda, where *1* is time of jet exposure on the surface being washed for 15 sec; *2* is time of jet exposure on the surface being washed for 45 sec; *3* is time of jet exposure on the surface being washed for 90 sec.

Data of this experiment corresponds to the following conditions of washing: diameter of the cap's cylindrical hole is 2 mm, pressure of the spraying jet is 2.5 kg/cm^3, temperature of the washing solution is 70°C, angle of jet direction to the can surface is 90°, distance between the cap and their surface is 200 mm. Increase in temperature of the washing solution leads to keen intensification of washing only to a certain limit that is to 70÷75°C. Its further increase is inappropriate and leads to unnecessary heat loss

It was established that the cleaned surface becomes larger with increase in pressure of the given solution. From the diagram (Figure 2) it is seen that the cleaned surface becomes smaller due to the solution pressure over 3 kg/cm^2. That is why it is inappropriate to increase pressure of the given solution for more than 3 kg/cm^2 for the tin cans with inconsiderable and easy-to-wash stains.

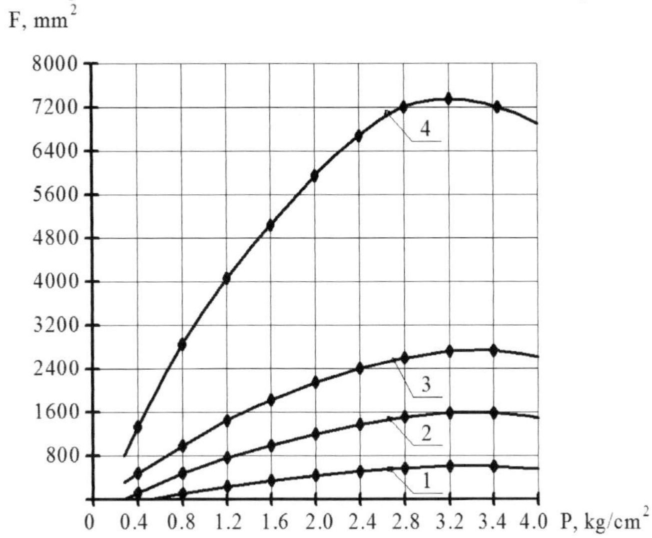

Figure 2. Dependence of the cleaned surface *F* on pressure *P* of the given solution with the diameter of the water jets: 1 mm-the 1st line; 2 mm - the 2nd line; 3 mm – the 3rd line; 5 mm - the 4th line.

Angle of jet direction from the cap to the can surface is also very important.

Efficiency of washing rises with angle increase (Figure 3) and reaches the maximum due to φ=90° and it decreases with angle decrease.

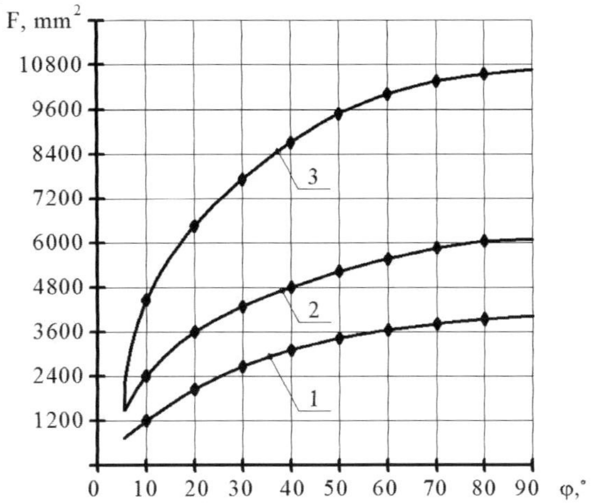

Figure 3. Dependence of cleaned surface on angle of φ - jet direction to the can surface with the diameter of cap's water jets, where the 1st line - 1 mm; the 2nd line - 2 mm; the 3rd line - 3 mm.

When distance between the cap and the can surface becomes larger (Figure 4), the cleaned surface decreases. Research shows that it is better to recommend the distance between the cap and the can surface of 90÷120 mm depending on position of collectors with caps in the washing machine to continuously moving cans and way of moving during washing and cans arrangement.

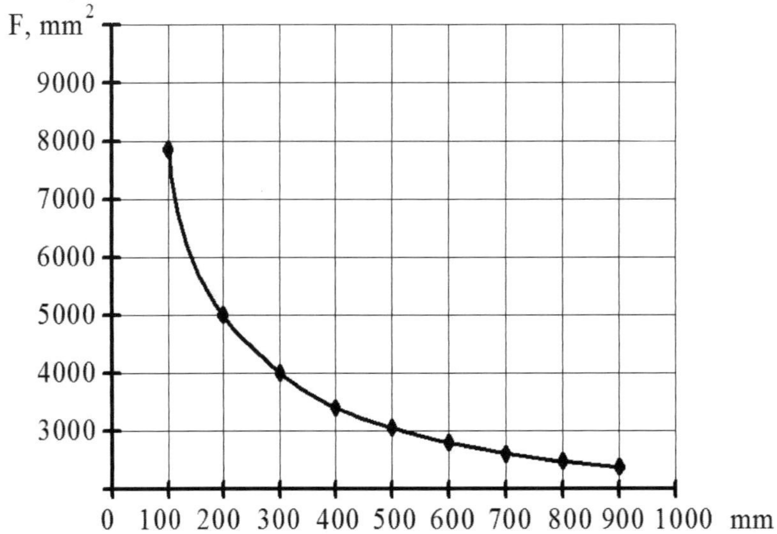

Figure 4. Dependence of cleaned surface on distance *l* between the cap and the can surface

4. Conclusion

Cleaned surface becomes larger with increase of concentration and temperature of washing solution.

Increase of solution concentration for more than 3% and its temperature for more than 70÷75°C do not promote intensification of the process.

Dependence of washing efficiency on solution pressure given to the cap with different diameters of its holes was found.

Furthermore, it is established that the process efficiency decreases with increase of pressure for more than 3 kg/cm^2, as a certain hydrodynamic mode of jet percussion action begins at the boundary with the surface being cleaned.

Cleaned surface depends on angle of jet direction at the point of its strike with the can surface. This dependence represents the fact that the cap should be placed in relation to transported cans in such a way that the jet would be always at right angle to the dirty surface.

The distance from the syringe nozzle to the can should be chosen within 90÷120 mm.

References

[1] Rauf S *et al* 2016 *Procedia Comput Sci* **83** 902–906
[2] Abdel-Fatah M A *et al* 2017 *Ain Shams Engineering Journal* **8(3)** 305–310
[3] Wahad H S *et al* 2017 *IOP Conf. Ser.: Mater. Sci. Eng.* **174** 012021 doi: 10.1088/1757-899X/174/1/012021
[4] Jain S V *and* Patel R N 2014 *Renew Sust Energ Rev* **30** 841–868
[5] Akikur R K *et al* 2013 *Renew Sust Energ Rev* **27** 738-752
[6] Plappally A K *and* Lienhard V J H 2012 *Renew Sustain Energy Rev* **16(7)** 4818–4848
[7] Tadesse M G *et al* 2017 *Smart Mater Struct* **26** 065016

IOP Conf. Series: Materials Science and Engineering **457** (2018) 012002 doi:10.1088/1757-899X/457/1/012002

Influence of the parameters of support wheels on the amount of the device of the trailer from the tracking course

S S Alatyrev[1], R V Andreev[2], A O Vasiliev[2], N N Pushkarenko[2], A O Grigoryev[2], S V Tikhonov[3]

[1]Professor, Chuvash State Agricultural Academy, Cheboksary, Russia
[2]Associate Professor, Chuvash State Agricultural Academy, Cheboksary, Russia
[3]Associate Professor, I. Ulianov Chuvash State University, Cheboksary, Russia
E-mail: 3777222@bk.ru

Abstract. The effect of the parameters of the left support wheel of the trailed cabbage machine, the hardness of the soil and its difference under the support wheels on the magnitude of the deviation of the unit from the direction of motion were studied experimentally. It is established that the rational values of the structural parameters of the supporting-running system of the trailed agricultural machine allow minimizing the values of the longitudinal vibrations of the working bodies during the execution of the technological process. Rotating parts of the trailing machine received movement from the power take-off shaft through a specially designed drive shaft. The system for measuring and recording the deviations of the machine from a given course is represented by a bridge electrical system based on a rheostat with a sliding brush connected via an analog port to an electronic data processing system, and then to a laptop computer. As a result of the research, it was found that with a decrease in the hardness of the soil under the right wheel, the stability of the stroke is sharply reduced, while the angle of deviation of the drawbar increases to unacceptable values. This is due to the fact that under these conditions, the resistance of the soil to the support wheels of the trailing machine is mainly applied on the right side, in addition, with decreasing hardness of the soil, the rolling resistance of the wheel increases. The results of the conducted studies confirm and supplement the results of theoretical studies. Investigating the stability of the movement of the trailer with support wheels of different sizes with the same hardness of the soil, it was found that the value of the angle of deflection of the machine's tail significantly decreases compared to the results of previous experiments and does not exceed the limits set by the agricultural requirements.

1. Introduction

Agriculture at all times is the foundation of the economy of any country, and ensuring the food security of its population is an important part of the economic policy of their governments [1]. There is Food Security Doctrine of the Russian Federation, adopted in 2010 and adjusted in 2014 [2].

In the second half of the last century, in our country and abroad, there was a trend of specialization in vegetable growing, enlarging the areas under the main vegetable crops. Conservation agriculture is one of the so-called emerging agrosciences [3] and encompasses techniques that minimize or eliminate tillage and, thus, maintain a vegetative cover that protects soil from its degradation. As a result of this concentration of production at the time, it was felt necessary to create more productive agricultural machinery. Certainly, conservation agriculture concept goes beyond conservation tillage and is defined by three linked core principles that must be jointly applied to create synergies [4]: minimum soil disturbance; permanent organic soil cover; and crop rotations. The results showed that new tillage did not affect grain yields and reduced 80% of energy consumption [5]. Qualitative indicators of the operation of trailed agricultural machines are in direct dependence on the stability of their movement in the aisle, the greatest influence on which is provided by natural conditions. These trials, which began in 1982 and still continue today, resulted in higher yields in new tillage fields than in those with traditional tillage [6].

Therefore, increasing the stability of the agricultural machine in the aisle is an urgent task in creating promising agricultural machines.

Earlier, on the example of a cabbage harvesting machine for determining the optimum parameters of a support-trailer part, a dependence was obtained:

$$Q = \frac{Pc}{r\sqrt{1 - \left(\dfrac{P}{0.5\pi brV}\right)^2}}, \tag{1}$$

where Q is the rolling resistance of the support wheel, N; P is normal load on the wheel, N; c is the rolling friction coefficient of the support wheel; r - the radius of the support wheel, m; b is width of the support wheel, m; V is hardness of soil, N / m^2.

For the existing machine sample, according to the Equation (1), it was established that the following dimensions of the support wheels should ensure the smallest deviations of the unit from the direction of travel:

- diameter of the left support wheel 620 mm;
- width of the left support wheel 145 mm;
- diameter of the right support wheel 762 mm;
- width of the right support wheel 235 mm.

In some studies, it was remarkable the collaboration with the technical departments of the industry [7]. All too often experimental protocols are not rigorous enough to unveil real differences in tillage systems [8–10].

The purpose of the research is to experimentally test the theoretically rational values of the constructive parameters of the support-running system of a trailed agricultural machine, which allow minimizing the values of the longitudinal vibrations of the working bodies during the execution of the technological process.

2. Experimental

In order to carry out an experimental test of the theoretical positions put forward, a laboratory field installation was built that can be aggregated with a tractor of class 1.4 (in this case MTZ-82) by means of a trailing drawbar. The rotating parts receive movement from the power take-off shaft through a specially designed drive shaft. The measurement system as well as the registration of the deviations of the machine from the given course is represented by a bridge electrical system based on a rheostat with a sliding brush connected via an analog port to an electronic data processing system (in this case - 3B NET log U11300ip), and then to a laptop computer.

Studies of the stability of the trailer were carried out on the plot of the areas with a differentiated hardness of the soil under the support wheels. The numbering of the areas is shown in Figure 1.

In this system, the resistance of the rheostat varies smoothly depending on the angle of the rotation of the drawbar of the trailing machine relative to the tractor's ear with the help of a sliding contact kinematically connected by a flexible connection with an arc rigidly attached to the drawbar of the machine centered at the trailer's point. During the movement of the installation, the resistance created by the rheostat changes the potential difference at the output of the bridge measuring circuit, that is, the value of the measured voltage is proportional to the angle of rotation of the drawbar relative to the trailer point.

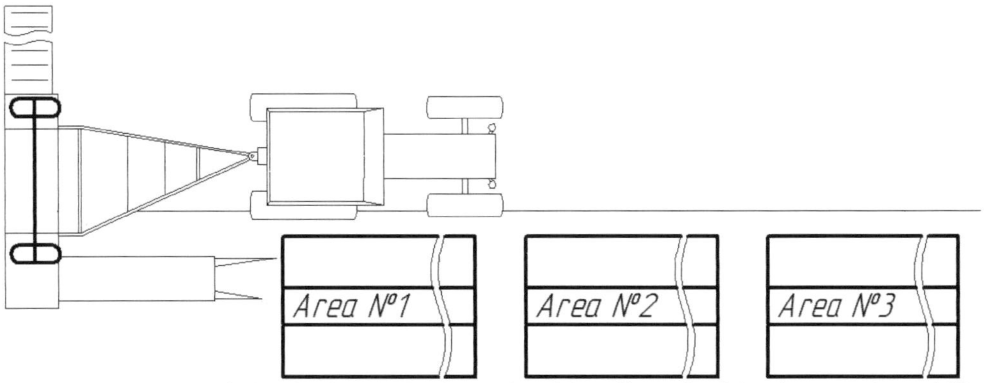

Figure 1. The number of plots in the plot in the study of the effect of soil hardness on the stability of the machine.

3. Results and Considerations

As it was already revealed theoretically, the condition for the best functioning of the trailer is to ensure the following of its working body strictly in the direction of movement of the unit. Compliance with this condition under the proposed support-trailer system was checked in the field by studying the machine's follow-up to a given straight-line course at previously prepared areas of the soil cover (Figure 1).

The investigations were carried out at predetermined parameters of the support-trailer system, ie: the length of the trailing drawbar is 2.35 m, the diameter of the right wheel is 762 mm and the width of the right wheel is 235 mm.

Variable parameters were the diameter and width of the left wheel, as well as the density of the soil in the sections.

Figure 2 shows a fragment of an oscillogram of the deviation angle of the unit from a given heading, recorded in real time. The oscillogram records the values of the deflection angle φ from the direction of travel of the trailer in the horizontal plane.

The received oscillograms allowed to analyze the process of trailer movement in the field plane in real time mode. It can be seen from them that on a plowed section of L_1-length soil the machine deviates from the set course (section T_1), and when entering a site with hard soil, the length L_2 approaches it (section T_2). Initially, $L_1 >> L_2$ is selected. The soil in each plowed area had different relative humidity W and, accordingly, hardness V. It is noticeable that with decreasing soil hardness the amount of deviation of the trailer from a given course increases somewhat.

For the sample inside the section T_1, it is possible to isolate the steady-state movement - T', on which the unit moves steadily after a transition from one site to another.

As the statistical analysis shows, the values of the angle of deviation of the trailing drawbar φ from a given course are predominantly subject to the normal distribution law, as evidenced by the results of estimating the discrepancies between the empirical and theoretical frequencies of the Pearson's variational series.

The nature of the variation in the angle of inclination of the trailer with the support wheels of the same (a) and different (b) sizes from the set course, depending on the position of the wheel (1-over the plowed plot, 2-over the unplowed plot) was considered with a straight trajectory of the machine.

The experimentally obtained values of the deviations of the trailing drawbar of the machine from the direction of motion are given in Table 1.

Table 1. The value of the angle φ deviation of the trailing drawbar from the direction of motion of the unit, deg.

Position of machine	The value of the deflection angle φ, deg. At $V = 168$ N/cm^2	At $V = 113$ N/cm^2	At $V = 90$ N/cm^2	Average value of angle, deg.
1-a	5.86	7.99	10.45	8.04
1-a	5.12	6.17	9.93	7.07
1-b	4.01	4.31	4.78	4.36
1-b	3.56	3.89	4.44	3.96
2-a	3.14	4.04	4.33	3.84
2-a	3.16	3.76	4.15	3.69
2-b	1.83	1.64	1.63	1.70
2-b	1.99	2.26	1.88	2.04

From Table 1 it can be clearly seen that with the reduction of the hardness of the soil in this position (under the right wheel) the stability of the stroke sharply decreases, while the angle of deviation of the drawbar increases to values exceeding in 2...3 times permissible for agricultural requirements. This is due to the fact that under these conditions, the resistance of the soil to the support wheels of the trailing machine is mainly applied on the right side, in addition, with decreasing hardness of the soil, the rolling resistance of the wheel increases.

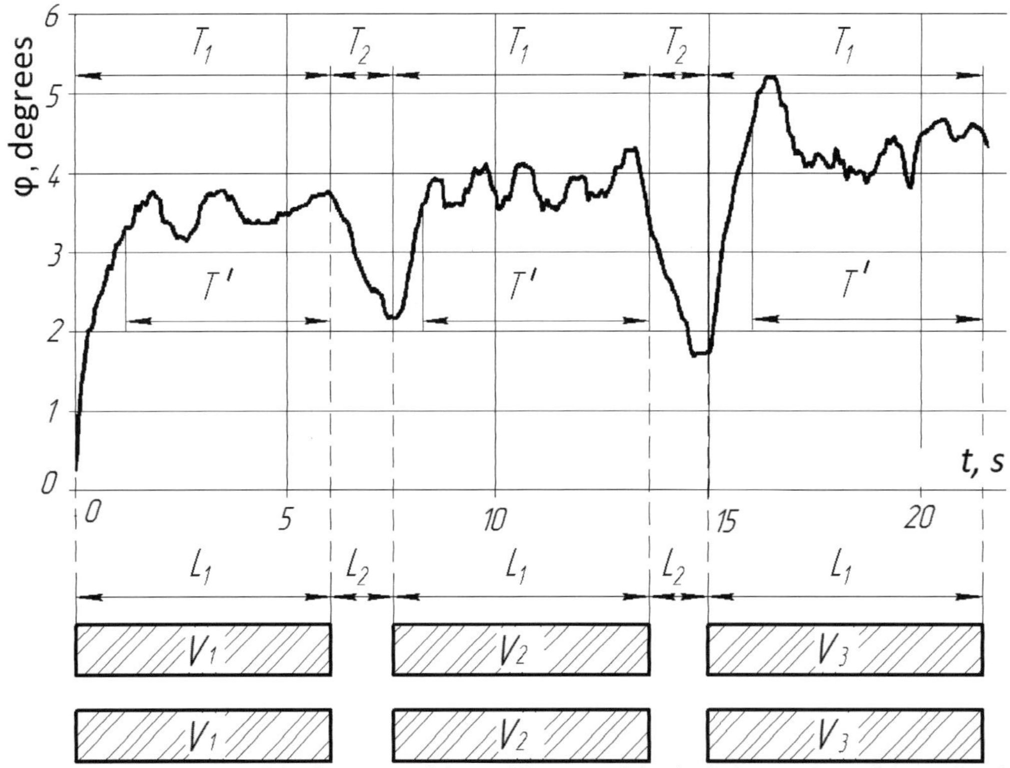

Figure 2. Component of the fragment of the oscillogram of the deviation angle φ of the trailer from a given course in different parts of the soil in real time t: L_1, L_2 – length of plowed and unplowed plots, respectively; T_1, T_2 – time of passage of the plowed and not plowed sections accordingly; T' is the time of steady movement of the aggregate along the plowed section; V_1, V_2, V_3 – the hardness of the soil.

Thus, the results of the carried out studies of the oscillations of the trailer machine in the horizontal plane are correlated and, complementing each other, confirm the results of theoretical studies.

Investigating the stability of the movement of the trailer with support wheels of different sizes in position 2, we note that if the soil hardness is close to the value under both wheels, the value of the drawbar angle significantly decreases compared to the results of the previous experiments and does not exceed the values set by the agricultural requirements.

4. Conclusion

Having studied the physical and mechanical properties of the soil as a medium with which the trailer machine interacts, the nature and degree of influence of the main factors of the process and the design of the support system on the stability of its movement in the aisle are theoretically and experimentally determined. It has been experimentally established that with the implementation of the radius and width of the left supporting wheel of a smaller size, and also the displacement of the trailer point to the right from the center by one hole, the amount of deviation of the trailed drawbar from the predetermined heading decreases to 1...3 degrees, which corresponds to 20...60 mm deviation of the cutting apparatus from a row of cabbage. This value does not exceed the established agricultural requirements.

It has been experimentally established that the theoretical values of the parameters of the support-trailer system (the diameter of the left support wheel is 0.62 m, the width of the left support wheel is 0.145 m, the diameter of the right support wheel is 0.762 m, the width of the right support wheel is 0.235 m, the length of the trailing drawbar is at least 1.44 m) are indeed optimal and rational for the performance of the work process by a trailer.

References

[1] Vasiliev AO *et al* 2017 *Proc of the 30th IBIMA Conference* 5294-5289
[2] Lozhkin AG *et al* 2017 *Proc of the 30th IBIMA Conference* 5295-5299
[3] Lichtfouse E *et al* 2010 *Agronomy for Sustainable Development* **30(1)** 1–1
[4] Kassam A *et al* 2012 *Field Crop Research* **132** 7–17
[5] Juste F *et al* 1981 *Proc of the 13 Conferencia Internacional de Mecanización* 133–145 (in Spanish)
[6] González Sánchez E-J *et al* 2010 *Vida Rural* **312** 24–27
[7] Fernández Quintanilla C 1997 *Agricultura de Conservación: Fundamentos Agronómicos, Medioambientales y Económicos. Asociación Española Laboreo de Conservación/Suelos Vivos* 1–12
[8] Karlen D L *et al* 1994 *Soil Tillage Research* **32** 313–327
[9] Calegari A *et al* 2008 *Agronomy Journal* **100(4)** 1013–1019
[10] Christopher S F *et al* 2009 *Soil Sci Soc Am J* **73** 207–216

IOP Conf. Series: Materials Science and Engineering **457** (2018) 012003 doi:10.1088/1757-899X/457/1/012003

Decision of objective object type recognition on the basis of using the diagram of antenna direction as a sign

S Boev[1], V Liferenko[2], S Kulikov[3], V Aldokhina[4]

[1]General Designer, JSC "RTI", Moscow, Russia
[2]Professor, A.F. Mozhaysky Military-Space Academy, Saint-Petersburg, Russia
[3]Senior lecturer, A.F. Mozhaysky Military-Space Academy, Saint-Petersburg, Russia
[4]Associate Professor, A.F. Mozhaysky Military-Space Academy, Saint-Petersburg, Russia
E-mail: YVikulova@oaorti.ru

Abstract. The paper presents an algorithm for recognizing the type of an object in aerospace on the basis of a comparison of directional patterns. Attraction of new features to the solution of the recognition problem will increase the probability of correct recognition and will allow the most complete description of recognized objects, which in turn will lead to a decrease in the probability of error in making a decision.

1. Introduction

In the Russian Federation, a complex spatially distributed system for collecting and processing information functions to monitor and control the activities of foreign states in the near-Earth space. One of the indicators of the quality of the work of this system is the probability of correct recognition. This indicator depends on many different factors and in some cases does not reach the level required by the guidance documents.

An increase in the indicator under consideration [1–3] with an unchanged number of types of recognized objects, it is necessary to increase the number of features in whose language these objects are described. This is achieved through the use of new measuring instruments, as well as the improvement of the scientific and methodological apparatus for processing measured information. In turn, as the analysis of recognition algorithms showed, the information coming from the complexes of radio technical control is practically not used in making the decision.

An analysis of the content model considered in [4–6] showed that for a typical recognizable object in the airspace, a certain number of modes of functioning are characteristic. Based on certain conservatism in the design of antenna systems and features of the operation of recognition objects, it is proposed in [7–9] to use as a complementary feature a spectral portrait and a diagram of direction.

This article presents an algorithm for recognizing the type of an object in aerospace space based on the use of the second of the above characteristics. The output data of the proposed algorithm will be used in a multi-level recognition system, where recognition is made on the first level based on various characteristics, on the second level - the computation of informativity and the assignment of weight coefficients, and, finally, the third decision is made on the basis of a set of characteristics using the "weighted voting" [1, 10, 11].

2. Research Method

2.1. Use of the distance function when recognizing on the basis of diagrams of direction

During the analysis of the directional pattern of antenna (DPA) in various modes of functioning of recognized objects [12, 13] it was established that, in contrast to spectral portraits, it is difficult to catalog the DPA sections. This is due, first of all, to the fact that the form of the real DPA has a complex multi-lobe structure (Figure 1). In this regard, it is advisable to use simulation simulation of the DPA to obtain its cross-section for these observation conditions. Simulation is performed for a certain type of recognizable object operating in a given mode of operation.

The decision on whether an object belongs to a particular type is made on the basis of the results of comparing the values of the cross sections of the DPA obtained in the course of the work session at the radio engineering complex and calculated in the course of simulation modeling.

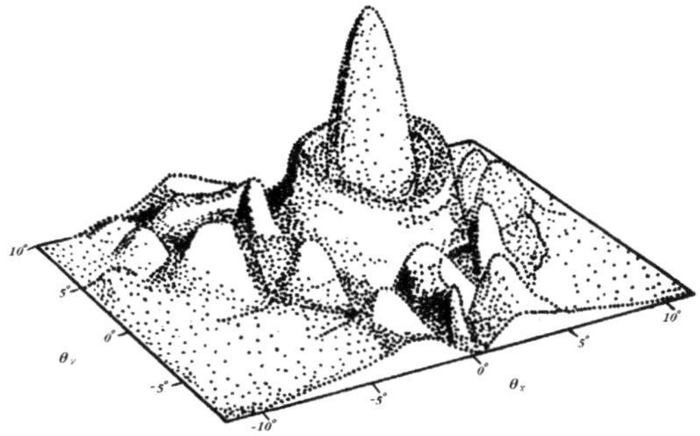

Figure 1. Directional pattern of a real antenna.

The principle of the minimum distance between the values of amplitudes, simulated and measured DPA is based on the above-described process. The difference in these values will show how much the measured DPA differs from the simulated one. At the same time, the cross section of the DPA is represented by a set of discrete samples, the size of which is advisable to be set as small as possible, which makes it possible to fully describe the entire "fine" structure of the DPA. Dependence of the amplitude difference of the simulated and measured DPA on the corresponding discrete will be called the difference portrait.

Consider the several difference portraits presented in Figure 2.

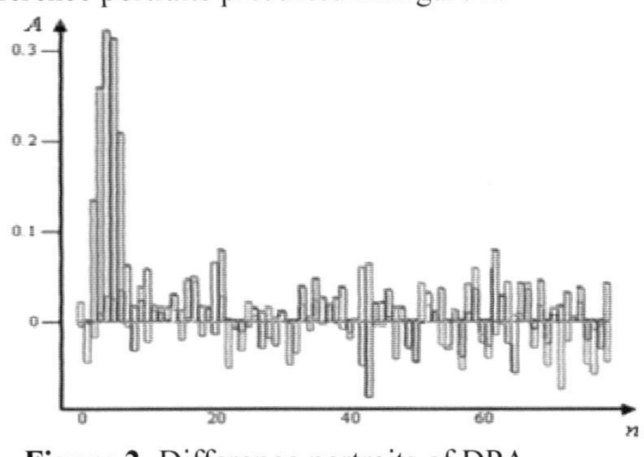

Figure 2. Difference portraits of DPA.

The first R_l^{izm} difference portrait reflects the degree of difference in the DPA obtained during the measurement session from DPA that obtained for the ideal case with the help of simulation modeling

$$R_l^{izm} = A_l - A_{il}^{imit.Id.}, \tag{1}$$

where A_l is a set of discrete, with the help of which the cross section of the DPA obtained during the measurement session is described; $A_{il}^{imit.Id.}$ is a set of discrete, with the help of which the cross section of the DPA obtained in the course of simulation is described; $i = 1,2,...,N$ is the number of object types; $l = 1,2,...,L$ is the number of disks with which the DPA is described.

The second R_{il}^{izm} difference portrait reflects the degree of difference between the two DPAs obtained with the help of simulation modeling, however, the values of one of them are corrected due to the influence of random factors, as if it were obtained in a real measurement session $R_{il}^{kat.} = A_{il}^{i\min.Sl.F.} - A_{il}^{imit.Id.}$. Evaluation of the influence of random factors should be carried out on the nearest to the recognized object, located in the zone of action of the complex of radio engineering measurements, the radiation parameters of which are known. Thus, the object can act as a "reference" object.

Comparing the difference portraits, one can assume that their mathematical expectations are equal only if the parameters of the DPA are identical. The small value of the discreteness of the difference portrait can be explained by the influence of external factors.

Thus, we propose the following hypothesis:

$$S_{ij} = \begin{cases} 1, & H_0 \text{ provided} \quad M(R_{il}^{izm}) \approx M(R_{il}^{kat}) \\ 0, & H_1 \text{ otherwise} \end{cases},$$

where H_0 is the hypothesis about the belonging of the recognized object to the *i*-type; H_0 is the hypothesis that the recognized object belongs to the *k*-type ($k \neq i$).

2.2. Application of the Criteria of the Student for Solving the Objective Type Recognition Object Type
In order to carry out object type recognition based on the antenna pattern, we will use the Z criterion (Student's criterion) considered in [6, 15]. In this case, when testing the hypothesis put forward, we will use the statistical method for correctness.

The principle of verification of statistical hypotheses is formulated as follows: if the observed value of the Z criterion belongs to the critical region z_{kr}, then the hypothesis is rejected if the observed value of the criterion belongs to the domain of hypothesis acceptance - the hypothesis is accepted [6, 16]. We divide the entire set of possible values of the criterion into three disjoint sets and form a two-sided domain defined by the inequality $|Z| > z_{kr}$, where z_{kr} is a positive number (Figure 3). In the figure, the case is considered when the critical points are symmetric with respect to zero under the assumption that $z_{kr} > 0$.

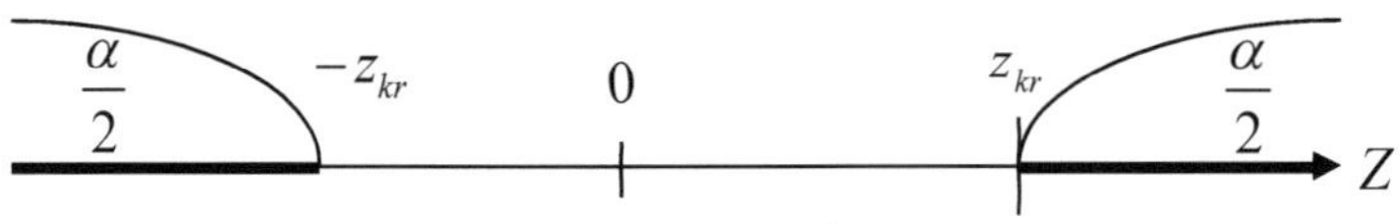

Figure 3. Two-sided critical region.

Thus, it is necessary to find a value $z_{кр}$ that would divide the set of values into three areas indicated above.

Analysis of the difference portraits showed that the values of the samples are random variables and as the number of measurements increases, the law of their distribution tends to normal. In their essence, the difference portraits are general samples of the amplitude values.

In turn, the sample means are unbiased estimates of the general average, i.e. $M(R_{il}) = M(\overline{R}_{il})$. In view of the fact that the general samples R_{il} are normally distributed, then the sample $\overline{R}_{il}$ means are also distributed normally.

Based on the foregoing, the quantity Z is a normalized random variable that is normally distributed, since it is a linear combination of normally distributed quantities $\overline{R}_{il}^{izm}$ and $\overline{R}_{iq}^{kat}$.

Based on the sample mean difference portraits R_{il}^{izm} and R_{iq}^{kat}, the value Z is calculated according to the following Equation

$$Z = \frac{\overline{R}_{il}^{izm} - \overline{R}_{iq}^{kat}}{\sigma(\overline{R}_{il}^{izm} - \overline{R}_{iq}^{kat})} = \frac{\overline{R}_{il}^{izm} - \overline{R}_{iq}^{kat}}{\sqrt{D(R_{il}^{izm})/L_{vyb} + D(R_{iq}^{kat})/Q_{vyb}}}, \tag{2}$$

where L_{vyb} is the number of discrete differences obtained when compared with the measured cross-section of the DPA; Q_{vyb} is the number of discrepancies of the difference portrait obtained in comparison with the simulated cross section of the DPA.

By the definition of the standard deviation - $\sigma(\overline{R}_{il}^{izm} - \overline{R}_{iq}^{kat}) = \sqrt{D(\overline{R}_{il}^{izm} - \overline{R}_{iq}^{kat})}$.

On the basis of the dispersion property of the difference of two independent random variables, we obtain $D(\overline{R}_{il}^{izm} - \overline{R}_{iq}^{kat}) = D(\overline{R}_{il}^{izm}) + D(\overline{R}_{iq}^{kat})$.

In connection with the fact that the variance of the arithmetic mean l of equally distributed mutually independent random variables is L times less than the variance D of each of the quantities, then: $D(\overline{R}_{il}^{izm}) = D(R_{il}^{izm})/L_{vyb}$, $D(\overline{R}_{iq}^{kat}) = D(R_{iq}^{kat})/Q_{vyb}$.

Consequently,

$$\sigma(\overline{R}_{il}^{izm} - \overline{R}_{iq}^{kat}) = \sqrt{D(R_{il}^{izm})/L_{vyb} + D(R_{iq}^{kat})/Q_{vyb}},$$

$l = q$ and $L_{vyb} = Q_{vyb} = L$ (the sample size is the same), therefore the Equation will take the following form

$$Z = \frac{R_{il}^{izm} - R_{il}^{kat}}{\sigma(\overline{R}_{il}^{izm} - \overline{R}_{il}^{kat})} = \frac{(R_{il}^{izm} - R_{il}^{kat})\sqrt{L}}{\sqrt{D(R_{il}^{izm}) + D(R_{il}^{kat})}}. \tag{3}$$

The next step is to determine the value of the criterial value z_{kr}, i.e., the boundary of the right-sided critical region based on a given level of significance α. To do this, it is necessary to find the value of the argument of the Laplace function, to which the value of the function equals $(1-\alpha)/2$. Then, in accordance with the foregoing, the domain of accepting the null hypothesis is determined by the inequality $|Z_{nabl}| < z_{kr}$.

Thus, when recognizing different types of objects and their belonging, it is proposed to use the following algorithm as additional features (Figure 4).

3. Results and Discussion

The proposed algorithm was used in recognition of the type of various objects [5], while the following results were given in Table 1.

Analyzing the data it can be concluded that the solution of the problem of object type recognition using a directivity diagram as a sign is possible. The limits of applicability of the considered algorithm are determined based on the requirements to the quality indicators of the recognition system.

Table 1. Results of solving the recognition problem using the Laplace function for different significance levels

$\varphi_1 - \varphi_2$	Number of correct hypotheses for different significance levels α (500 experiments), [%]				
	0.01	0.05	0.075	0.1	0.15
0.75 – 0.75	99	94.2	91.4	88	81.6
0.75 – 0.9	5	20.4	28.2	36.6	46.4
0.75 – 1.0	13.6	52.8	64.8	72.4	82.6
0.75 – 1.25	50.8	93.8	97.6	99	100
0.75 – 1.5	91.2	100	100	100	100
0.75 – 2.0	100	100	100	100	100

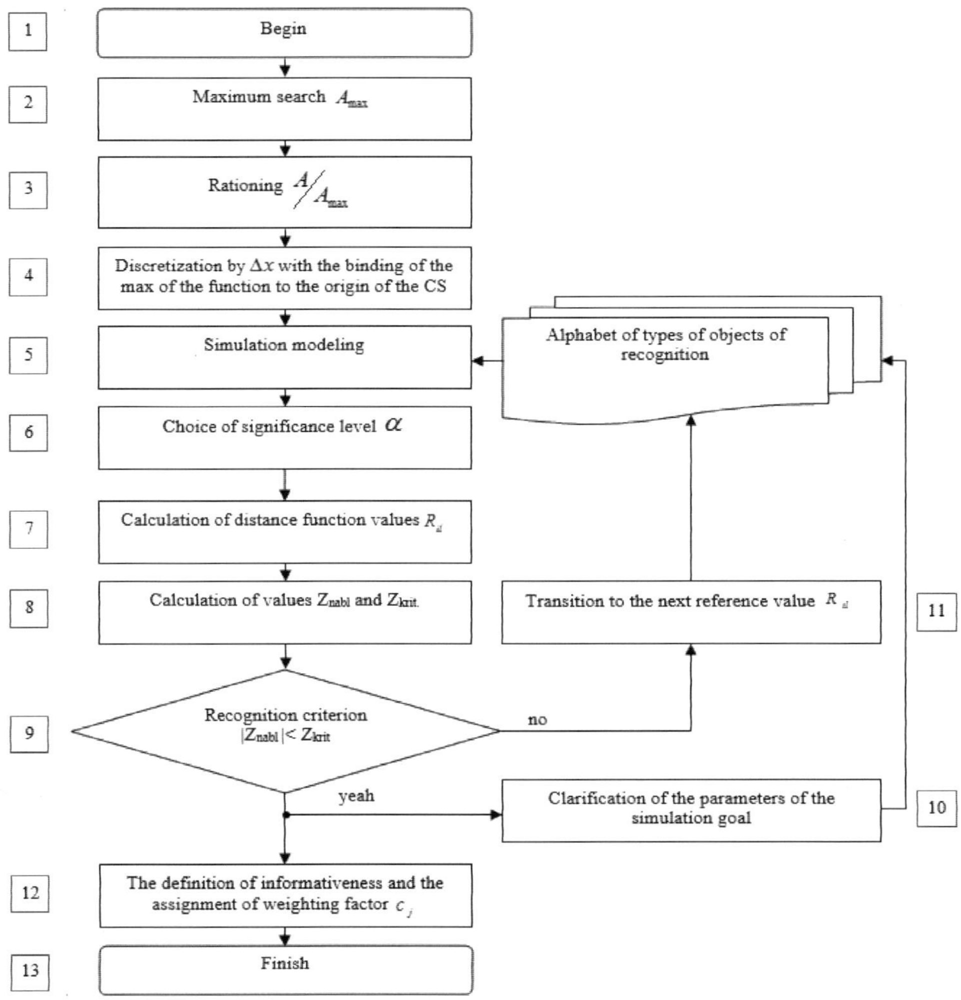

Figure 4. Algorithm for recognizing the type of the object based on directional patterns.

4. Conclusion

The considered algorithm allows to carry out the recognition of the type of the object on the basis of the comparison of the directional patterns with the possibility of variation of the threshold values due to the change in the value of the significance level. The recognition solution, obtained with the help of this algorithm, is one of the elements of the initial data for the realization of the multichannel recognition system with discrete accumulation.

References

[1] Gorelik A L *et al* 1990 *Selection and recognition based on location information* (Radio and Communication: Moscow)

[2] Randen H J V *et al* 2016 *Einführung in UML: Analyse und Entwurf von Software* (Springer Vieweg: Wiesbaden)

[3] *Global Climate Observing System* Retrieved from https://public.wmo.int/en/programmes/global-climate-observing-system

[4] Gudaev R A *et al* 2015 *Proc of the Military Space Academy of A.F. Mozhaysky* **649** 159–167

[5] Rincon R F *et al* 2011 *IEEE Trans Geosci Remote Sens* **49(10)** 3622–3628

[6] *Fraction of Absorbed Photosynthetically Active Radiation* Retrieved from http://fapar.jrc.ec.europa.eu/WWW/Data/Pages/FAPAR_Download/FAPAR_Download.php

[7] Vasin V V *and* Stepanov B M 1977 *Reference book on radar* (Soviet radio: Moscow)

[8] Seidl M *et al* 2015 *UML @ Classroom: An Introduction to Object-Oriented Modeling* (Springer International Publishing: Switzerland)

[9] Rast M *et al* 1999 *Int J Remote Sens* **20(9)** 1681-1702 doi: 10.1080/014311699212416.

[10] Krieger G *et al* 2008 *IEEE Trans Geosci Remote Sens* **46(1)** 31–46.

[11] Pinty B *et al* 2002 *A time composite algorithm for FAPAR products: theoretical basis document* (Institute for Environment and Sustainability: Italy).

[12] Lievers C M *et al* 2007 *Proc of EuRAD* 43–46 doi: 10.1109/EURAD.2007.4404932.

[13] Osis J *and* Donins U 2017 *Topological UML Modeling: An Improved Approach for Domain Modeling and Software Development* (Elsevier: US).

[14] Younis M *et al* 2003 *IEEE Trans Geosci Remote Sens* **41(71)** 1735–1739.

[15] Hooker S B *et al* (eds) 1992 *SeaWiFS Technical Report Series* **1** 104556.

[16] Bevilacqua A 2014 *Redmine plugin extension and development: build stunning extensions quickly and efficiently by leveraging Redmine's plugin facilities* (Packt Publishing: UK).

Effectiveness of effect of antioxidant of new generation on oxidizing processes in the containing collagen raw materials of the poultry-processing industry

N N Kuzmina[1], O Yu Petrov[2], E A Savinkova[2], I V Guseva[1]

[1]Udergraduate, Mari State University, Yoshkar–Ola, Russia
[2]Associate Professor, Mari State University, Yoshkar–Ola, Russia
E-mail: kuzmina221995@mail.ru

Abstract: The purpose of researches is comparative assessment of effectiveness of effect of natural antioxidant at storage of raw materials of the poultry processing industry. For achievement of a goal addition of antioxidants is provided in control specimens. Addition of dihydroquercetin to the emulsified skin of broiler chickens, at a rate of 0.50; 0.75 and 1.00 kg per 100 kg of raw materials, the way to achieve a significant level of their oxidative damage. The experimental data's are obtained as a result of carrying out laboratory researches on definition of acid and peroxide numbers in raw materials of the poultry processing industry in 7, 14, 21 and 28 days of storage. Antioxidant with various effectiveness promoted decrease in extent of their oxidizing decay. Addition of antioxidants in test pieces of raw materials promoted essential braking of its oxidizing decay. Their acid number over the period of the experiment was 1.3-1.7 times lower, and the peroxide number 1.8-3.6 times lower than in the control samples stored without the addition of an antioxidant. The received results convince of high performance of use of dihydroquercetin as the fissile antioxidant that provides a possibility of its use along with the available analogs.

1. Introduction

At deep processing of carcasses of broilers invaluable, in the food relation, products are recommended to be used mainly for preparation of soup kits and jelly which are not in great demand at the population and besides are not subject to the long-term storage. Many enterprises use these proteinaceous raw materials on development of dry feeds of animal origin. At the same time it should be noted a lack of theoretical researches and, as a result, validity of approaches in the solution of the modern technical tasks in the field of processing of secondary raw materials of the poultry-processing industry [1].

Though nutritionists do not recommend eating this type of raw materials because of the high content of fat, you should not forget about its useful properties. It is known that skin of broilers contains the significant amount of a collagen. At high concentrations in compounding of products from fowl a collagen can affect on the functional properties of myofibrillar proteins. A collagen can cause decrease of the sizes (contraction) of products from the crushed meat, especially at high-temperature processing and also can influence binding of pieces of meat in the formed products [2].

The specified properties are used in compounding of the new developed products from fowl which part skin of broilers is. The purpose of creation of such products is, first of all, expansion of the range of food and improvement of their quality [3].

When developing such products addition of antioxidant of dihydroquercetinum can provide the significant increase in duration of terms of their storage and will allow to give the functional orientation to products.

Recently considerably interest in phytomedicines as they are safe and more physiologic for a human body, than synthetic additives habitual to the modern medicine extended. Medicines of a phytogenesis are most widely presented by flavonoids. The most significant representative of this class of connections - dihydroquercetinum (DHQ). It is the fissile antioxidant, a unique natural acceptor of the free radicals, a gepatoprotektor and a radioprotector, the medicine having the anti-inflammatory and anesthetizing properties. Due to the DHQ high complexing properties brings out of an organism heavy metals, including radionuclides [4].

Dihydroquercetinum – the natural connection allocated from wood of the Siberian larch is recognized as one of the activest antioxidants of natural origin applied by production of foodstuff [5]. Digidrokvertsetin has the highest antioxidatic activity. Antioxidatic activity is the index reflecting ability to inactivate the free radicals of oxygen. The antioxidatic activity of dihydroquercetinum is shown at its concentration $10^{-4} \div 10^{-5}$. It is the least concentration of substance with antioxidatic activity in comparison with all known exogenesis antioxidants, including reproduction vitamins, A, B, C, D, K, β-Carotinum. Nevertheless, use of natural antioxidants by production of meat products, is caused by emergence of particular difficulties at stages of technological process. Vitamins antioxidants of natural origin easily lose the activity and their structure under the influence of various factors collapses: contact with a metal surface, influence of open sunshine and oxygen of air, high temperature processing (an exception vitamins A, E, K and carotenoids are) [6, 7]. With the researchers conducted in the Moscow Medical Academy of I.M. Sechenov it is confirmed that dihydroquercetinum is nontoxic, physiologically harmless to health of the person, does not give to products of strangers of smack and a smell, does not change their color at its use. Additive is steady in relation to temperature (from minus 50 to plus 180°C), to mechanical influences, and the processes taking place at manufacture of products that is meets all requirements imposed in general to all nutritional supplements and, in particular, to antioxidants. This connection is included in the list of the nutritional supplements which are not making harmful effects on health of the person when using for preparation of foodstuff (the SanPiN 2.3.2.1078-01 "Hygienic requirements to safety and nutrition value of foodstuff") [8].

And the main thing - dihydroquercetinum - is a unique immunomodulator. Taking into account the modern living conditions of people, dihydroquercetinum is the substance necessary for a general population as therapeutic means on already developed illnesses and for their prophylaxis as medicine which will allow keeping health and activity for many years.

Development of new products provides use of ecologically safe raw materials, and addition of dihydroquercetinum will provide increase in duration of terms of their storage by 2÷4 times, due to high antioxidatic activity and will increase biological value. It will allow giving the functional orientation to products and will create conditions for prophylaxis of a number of diseases, thanks to the maintenance of DHQ possessing to kapillyaroprotektorny, anti-inflammatory, radio tire-tread, disintoxication and hepatoprotective properties.

The release of these products will promote the quickest response to inquiries of consumers, updating of the range and its orientation, including to specialized groups of consumers. It is bound to specifics of the technology capable it is easy to modify process, to use secondary raw materials of the poultry-processing industry, ways of its preparation and application. Therefore production of products with use of secondary raw materials remains the most dynamically developing sector both on the outputs, and on the range and price categories.

2. Experimental

Laboratory researches on definition of the acid and peroxide numbers demonstrating formation of products of oxidizing decay of fat in skin from carcasses of broilers within 28 days of storage were conducted. As the experimental exemplars served raw materials in the cooled state.

According to a goal, objects of a research served:

- skin from the cooled meat of broilers of 1 grade in accordance with GOST P 52702-2006;
- the Dihydroquercetinum dietary supplement according to technical specifications 9100-241-21428156-11, according to the recommendations of the State sanitary and epidemiologic rationing of the Russian Federation No. 2.3.1.1915-04 of 2004 [8].

Medicine dihydroquercetinum was added to test pieces of the skin of broilers finely crushed, homogenized in number of 0.50; 0.75; and 1.00 kg on 100 kg of raw materials. In 7, 14, 21 and 28 days of storage of exemplars at a temperature of 3±1 °C, definition, in 3-fold frequency, contents in control and test pieces of oxidates on acid and peroxide numbers, according to the practical reference standards was carried out. The received results are processed by methods of mathematical statistics.

3. Results and considerations

The comparative analysis and complex assessment of content of oxidates in control and test pieces objective confirms effectiveness of the inhibiting effect of dihydroquercetin on intensity of oxidizing processes in the containing collagen raw materials of the poultry-processing industry.

Increase in acid number demonstrates education in raw materials of the free fatty acids which are formed as a result of hydrolytic decay of fats. In the studied exemplars of an emulsion of chicken skin, in 7 days of storage (Figure 1), this index the largest size had in a control specimen, free of antioxidant medicine – 0.30 mg the KOH/g which in 28 days of storage, naturally increased to 0.40 mg the KOH/g.

Addition of dihydroquercetinum in test pieces of raw materials promoted essential decrease in extent of its oxidizing decay. At introduction of antioxidant at the level of 0.50 kg / 100, in 7 days of storage, acid number was raw materials kg below monitoring for 33%, at addition of 0.75 kg / 100 raw materials kg, respectively, for 40%, and at concentration of 1.00 kg / 100 the raw materials kg – is 63% less than this index in a control specimen. Similarly, in 28 days of storage, acid number, respectively, was 25 lower; 45 and 70% on the studied concentration of DHQ.

In general, it was established that in 28 days of storage the exemplars containing the highest concentrations of DHQ were subject to the least oxidizing changes. Acid number during experience increased in them only by 9÷22%, against 33% in monitoring. At the same time, the value of this index in a control specimen reached the KOH/g level of 0.40 mg characterizing it in a condition of decay in 28 days of storage in the cooled state. At the same time the exemplars containing antioxidant in all studied concentration met the requirements of the specifications and technical documentation that testifies to its high performance.

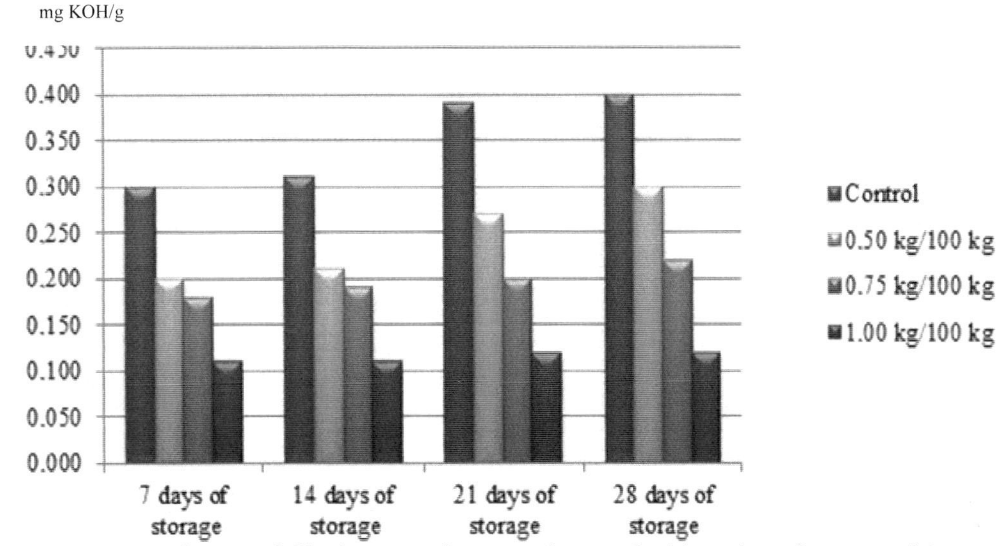

Figure 1. Influence of dihydroquercetinum on change of acid number of raw materials.

When studying antioxidatic activity of dihydroquercetinum, in parallel with acid number, peroxide value - the index characterizing amount of primary oxidates of lipids (hydroperoxides and peroxide compounds) in the containing collagen raw materials (Figure 2) was defined.

In a control specimen, is more narrow in 7 days of storage of raw materials, the value of peroxide value reached the indicator characterizing it as to the fresh, but not subject storage (0.020 mmol $(^1/_2O_2)$/kg) while the exemplars containing DHQ were estimated as fresh and liable to storage. At introduction of dihydroquercetinum to raw materials test pieces, its influence on peroxide value it was positive and objective. So, at addition of antioxidant in number of 0.50 kg / 100, in 7 days of storage, peroxide value in an exemplar was nearly 1.8 times lower than raw materials kg, in comparison with monitoring.

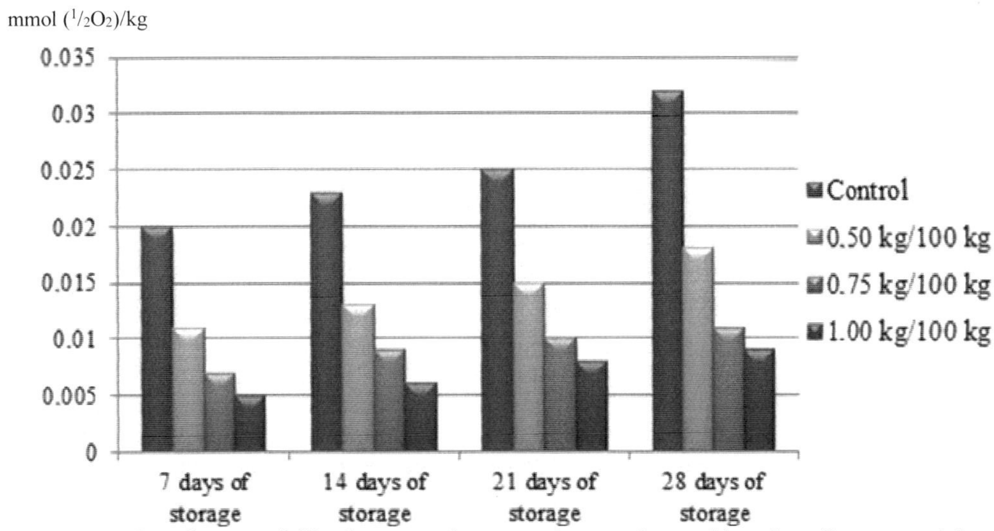

Figure 2. Influence of dihydroquercetinum on change of peroxide value of raw materials.

At the content of dihydroquercetin in exemplars at the level of 0.75 and 1.00 kg / 100 kg of the containing collagen raw materials, peroxide value in these exemplars was even lower – respectively in

2.9 and 4 times. And in 28 days of storage this index in test pieces relatively in 1.8 monitorings was lower; 2.9 and 3.6 times, respectively. For this period, the least intensity of accumulation of peroxide compounds is noted in a test piece with concentration of antioxidant 0.75 kg / 100 in kg of the containing collagen raw materials. The peroksidny number increased in it by 57%; it is 7% lower than intensity of accumulation of peroxides in the first test piece (0.50 kg / 100 kg) and 23% below than in a test piece with the highest concentration of antioxidant (1.00 kg / 100 kg).

4. Summary

The received results objective demonstrate efficiency of effect of antioxidant of dihydroquercetinum at the level of its concentration of 0.75 kg / 100 kg, in the containing collagen raw materials of the poultry-processing industry, on formation of oxidates at storage and allow to consider it optimum.

Therefore, dihydroquercetinum shows high antioxidatic activity, interfering with accumulation of oxidates, and promotes the significant increase in shelf-lifes of the containing collagen raw materials.

References

[1] Hellberg R S *et al* 2017 *Food Control* **80** 23–28 doi: 10.1016/j.foodcont.2017.04.025
[2] Karre L *et al* 2013 *Meat Sci* **94(2)** 220–227 doi: 10.1016/j.meatsci.2013.01.007
[3] Tomasevic I *and* Rajkovic A 2015 *Procedia Food Sci* **5** 285–288
[4] Petracci M *et al* 2013 *Trends Food Sci Technol* **33** 27–39
[5] Mani-López E *et al* 2012 *Food Res Int* **45(2)** 713–721
[6] Kalaba V *et al* 2017 *IOP Conf. Ser.: Earth Environ. Sci.* **85(1)** 012082 doi: 10.1088/1755-1315/85/1/012082
[7] Cassidy E S *et al* 2013 *Environ Res Lett* **8(3)** 034015 doi: 10.1088/1748-9326/8/3/034015
[8] Methodical recommendations of the State Sanitary and Epidemiologic Rationing of the Russian Federation no. 2.3.1.1915-04 2004 *The recommended consumption levels food and biologically the active materials* (in Russian)

Parameters of air flow over the upper sieve in the cleaning chamber

N V Yanukov[1], A V Mayorov[1], D A Mikheeva[2], A I Volkov[1], D V Lukina[1]

[1]Associate Professor, Mari State University, Yoshkar-Ola, Russia
[2]Graduate Student, Mari State University, Yoshkar-Ola, Russia
E-mail: ao_maiorov@mail.ru

Abstract. The value and direction of the resultant vector of the airflow coming out of the upper sieve, depending on changes in the gap between the blinds in purification of the combine harvester, is assessed. The volume and direction of the resultant vector of the airflow, as well as its components were evaluated in the study. Dynamic pressure was measured over a sieve with Pitot tubes in 6 points along the length, in each section at a distance of 50 mm from the blinds mounting axes. During the purification with a serial sieve by S_{vr} = 7...13 mm speed is almost constant, but when S_{vr} >13 mm, it intensively reduces from 4 to 2.50 m/s. This mentioned pattern of speed changes of serial sieves affects undoubtedly negatively on the separation of grain. In fact, by increasing of the supply of straw heap, S_{vr} gap between the louvers is increased to reduce static resistance of network. However the increase of S_{vr} reduces the air flow speed and its loosening impact on heap. Thus, one reason for intensive grain losses increase in empty glume by higher feed rates of heap of straw is a decrease in the function v (S_{vr}). Under the influence of the vertical component of velocity and vibrations of sieves the heap layer comes into a quasi-fluid state and the frictional force between particles decreases. It facilitates the passage of grains through the layer. During the tests in purification units of harvesters measurements of air flow velocity over the sieves were made.

1. Introduction

Cleaning of modern combine harvesters operate in conditions of higher loads and do not have enough reserve capacity to ensure minimal losses in the overtails from purification in any conditions of harvesting. Low throughput capacity of purification constrains further intensification of the process of threshing and separation of heap of straw, thus increasing the throughput capacity of the combine is needed [1–3].

The separation of grain in the process of cleaning of a grain harvester is influenced by many factors: the value of non-grain fraction feed, composition and content of straw impurities in the heap of straw, moisture of grain and non-grain fractions, the uniformity of flow, structural and kinematic parameters of purification, performance of the air system, and others [4].

The amount of feed for purification and composition of straw heap is determined by feeding of the plant mass into grain harvester thresher, its technological parameters, structural and kinematic parameters of the threshing and separating device (TSD).

Factors that influence the feeding amount of straw heap for cleaning are unmanageable (technological characteristics of the plant mass) or can not be changed (the feeding of plant mass into a TSD, the parameters of TSD) because of the desire to get maximum grain release from the straw and the best performance of threshing and separating device [5–7].

Separation of grain in screen-air cleaners depends much on their structural and kinematic parameters, changing which we can increase the intensity of grain separation from heap of straw and reduce losses.

2. Experimental

The volume and direction of the resultant vector of the airflow, as well as its components were evaluated in the study. Dynamic pressure was measured over a sieve with Pitot tubes in 6 points along the length, in each section at a distance of 50 mm from the blinds mounting axes.

The overall picture of the process for studied sieves is represented as follows.

Airflow with a serial sieve, leaving the fan nose, is divided with cross bar of the lower sieve into two unequal parts (Figure 1). The upper, smaller part of the air stream, skirting the bar, is directed at an angle of 75–80° to the plane of the lower flap of louvered comb of upper sieve and passes through 6–9 rows of combs. Another part of the air flow moves between the upper and lower sieves. At the same time attacking the lower flaps of the upper comb of the sieve the airflow partially goes through them. Turbulence is created and a layer of heap on the top sieve is blown ineffectively. The lower, bigger part of the flow, attacking the lower sieve, passes through the first 10÷12 rows of combs. After that at the length of 2/3 of the sieve between the combs intensive secondary eddy currents - «separated bubbles» are formed. It is connected with the location of lower claps (comb elements that are placed at a big angle to the main flow). A head-on impact and boundary layer separation happen. They reduce the air flow cross-section between the combs because of the formation of the vortex and the air flow through the sieve. This leads to a reduction of air flow speed, leaving a series sieve, in 1.5 time (from 2.30 to 4 m/s). An airflow loses its ability to blow through and loosen a thick layer of straw heap, thus causing increased losses for purification.

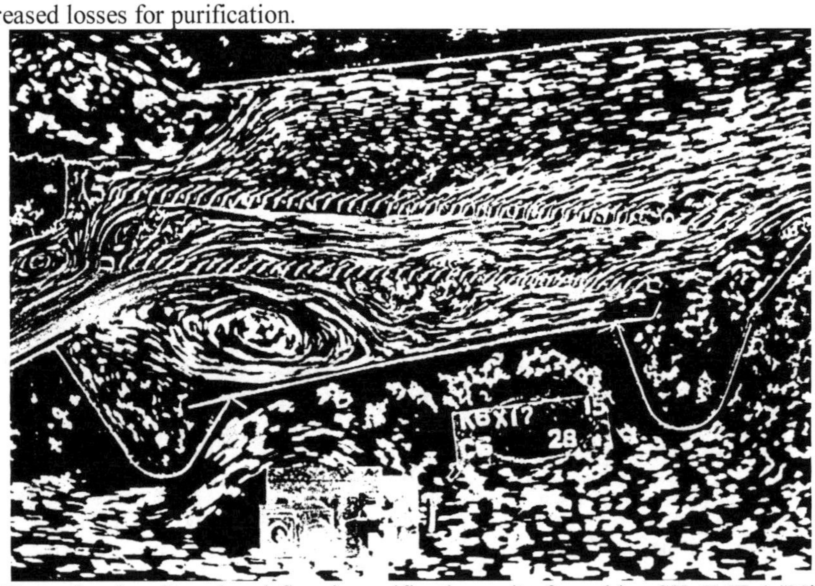

Figure 1. Flow of series sieves by airflow in purification unit of combine SK-5ME-1 "Niva-Effect".

In addition, the grain that was dressed through a layer and got on the surface of the louvers comb is thrown again by vortex current in a heap.

In experimental sieves the angle between the direction of the airflow and plane of the lower flap of the comb is small, so the punch of the air flow on them is eliminated. Dynamic pressure changes less than with serial sieves, eddy currents in the zone between the flaps are eliminated. Absence of the latter increases the speed of passage of grain in the space formed by flaps, and recurrent release of grain into the flow moving over sieves is decreased. Along with this, the direction of air flow, leaving the space between the flaps, varies immaterially depending on the position of the louvers. Its main vector is inclined to the plane of movement of the heap at a larger angle.

IOP Conf. Series: Materials Science and Engineering **457** (2018) 012005 doi:10.1088/1757-899X/457/1/012005

Figure 2 shows a diagram of the main directions of the main vector airflow over series (*a*) and experimental (*b*) sieves of purifications.

As can be seen from the figure, the direction of the principal vector in serial sieve purification varies in range β=35...60^0 with S_{vr} louver opening from 13 to 20 mm. In the experimental sieve the vector is diverted from the plane of the sieve at a larger angle β (65...90^0 by S_{vr}=13...20 mm). When S_{vr}≤10 mm, the angle of direction of the main vector in the experimental purification is 2 times bigger than in series one, i.e. 55 and 25^0.

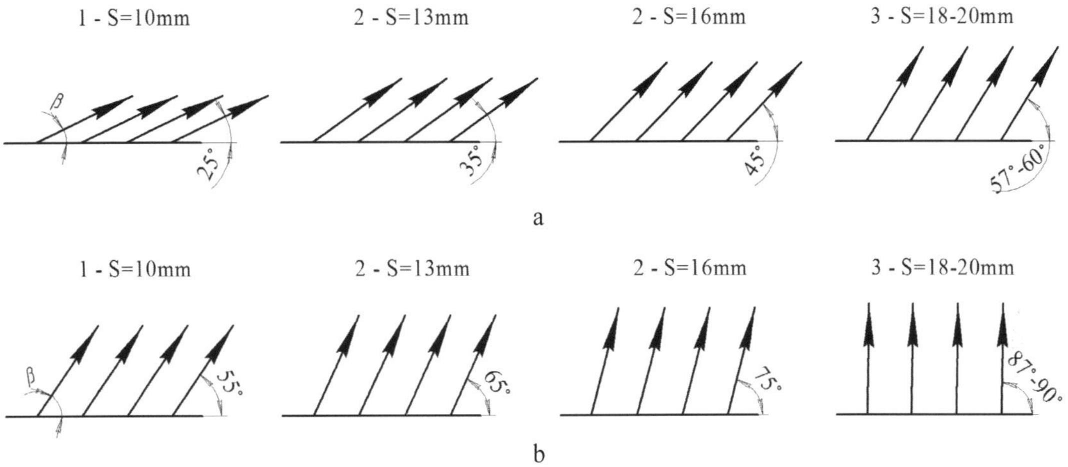

Figure 2. Changing the direction of the resultant vector of the air flow over the sieves of serial (*a*) and experimental (*b*) purification units.

The direction of the resultant vector affects greatly on the interaction of air flow with a heap: in the experimental purification (Figure 3) the air looses and fluffs a heap more intensive and contributes to its transportation along the sieve less. In the series sieves its transporting role, especially when S_{vr}≤13 mm, is essential.

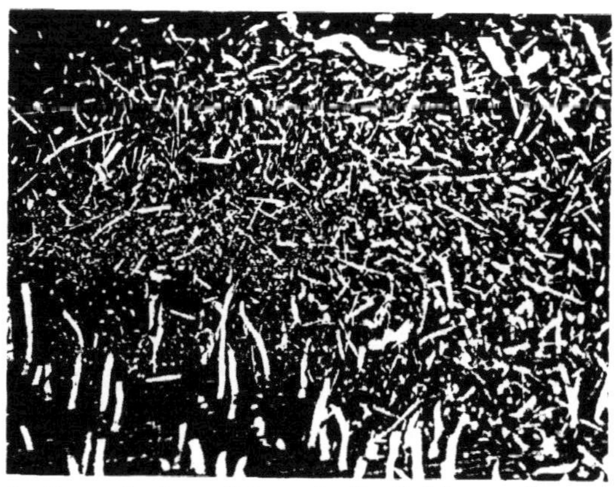

Figure 3. General view of the velocity field over the upper sieve.

It is characteristically noted that the average speed of the main vector of air flow in purification with experimental sieve changes (Figure 4) with increasing S_{vr} from 7 to 13 mm with a greater intensity than by S_{vr}>13 mm. When S_{vr} >13 mm, then v = 5 m / s.

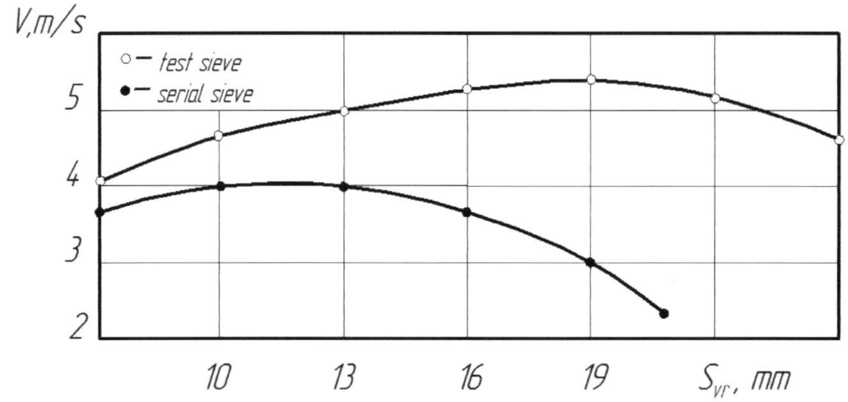

Figure 4. Change in the average speed of the main vector of the air flow over the sieves because of increase of the gap S_{vr} between the louvers of the upper sieve.

3. Results and considerations

During the purification with a serial sieve by S_{vr} = 7 ... 13 mm speed is almost constant, but when S_{vr} >13 mm, it intensively reduces from 4 to 2.50 m/s.

This mentioned pattern of speed changes of serial sieves affects undoubtedly negatively on the separation of grain. In fact, by increasing of the supply of straw heap, S_{vr} gap between the louvers is increased to reduce static resistance of network. However the increase of S_{vr} reduces the air flow speed and its loosening impact on heap. Thus, one reason for intensive grain losses increase in empty glume by higher feed rates of heap of straw is a decrease in the function v (S_{vr}).

Greater understanding of the regulation results gives the dependence of the results of vertical v_y and horizontal v_x components from the average speed from the opening of louvers (Figure 5). It should be noted that the horizontal component has an influence on the velocity of the heap of straw on sieves, and the vertical one has – on loosening effect. With an increase of the opening of series sieves blinds vertical velocity component varies slightly, and the horizontal is constant before the opening of louvers on 10 mm, and then gradually decreases. Vertical velocity component of the experimental sieves increases from 2.40 to 3.60 m/s, and the horizontal decreases from 2 to 1.40 m/s.

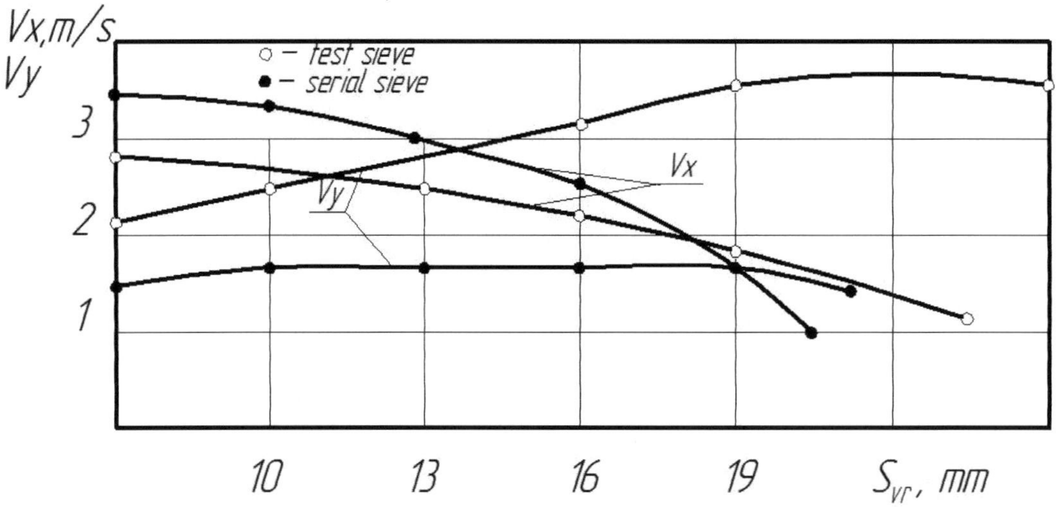

Figure 5. Influence of S_{vr} gap between the sieves' louvers on horizontal v_x and vertical v_y which are the components of the velocity of the flow over sieves.

Under the influence of the vertical component of velocity and vibrations of sieves the heap layer comes into a quasi-fluid state and the frictional force between particles decreases. It facilitates the passage of grains through the layer.

During the tests in purification units of harvesters measurements of air flow velocity over the sieves were made. The air flow speed over the sieves of experimental purification unit was measured in the gap S_{vr} between the louvers 22 mm and serial – by S_{vr} = 18 mm.

Figure 6 shows the velocity distribution diagram for air flow over sieves purification units.

In the Figure 6*b* it can be seen that the speed of serial purification decreases from 6 to 3 m/s at a distance of 200–300 mm from the beginning of the sieve, and then at the length of 1.20 m remains at the level of 3.50 m/s.

This is because of the situation when S_{vr} = 18 mm (65°) in the louvers space, decrease of circulation velocity and swirl of air flow happen. The air velocity at the exit of the sieves thus decreases, because of it the heap is not loosened enough. The main part of the air passes between the upper and lower sieves.

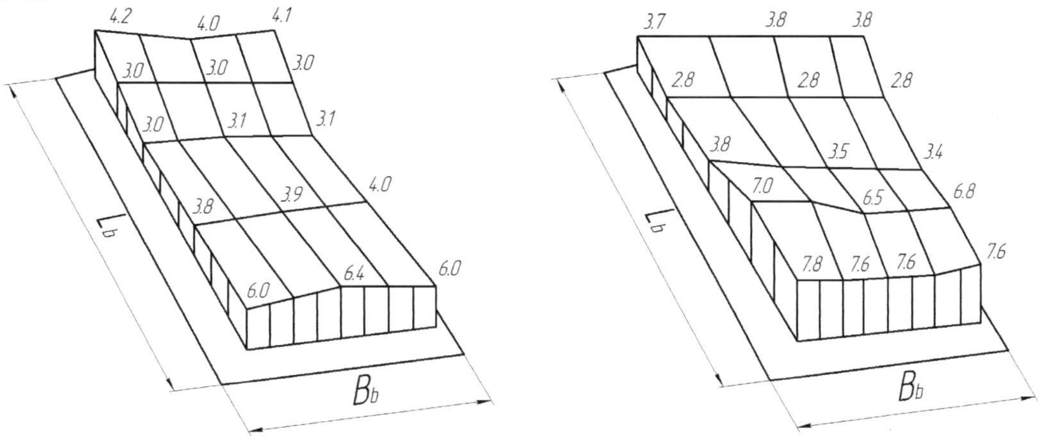

a b

Figure 6. Diagrams of distribution of the air flow speed over the sieves of the purification unit of combine SK-5ME-1 "Niva-Effect", where *a* – experimental, n=610 min^{-1}; S_{vr}=22 mm; *b* – serial, n=610 min^{-1}; S_{vr}=18 mm

In the experimental purification unit (Figure 6*b*) a smooth decrease in the speed of the air flow from the beginning to the end of the sieve (from 7.60 to 3.50 m/s) happens, so the heap layers are sufficiently loosened. The grain losses are reduced in comparison with the serial purification unit.

4. Summary

It follows from the foregoing that with the increase of the gap S_{vr} between the combs of experimental sieves the air flow velocity increases in the output. This increases the vertical velocity component, whereby a heap of straw, being in a suspended state, is actively purged and loosened.

Increasing of S_{vr} between combs of serial sieves results in a reduction of the speed at the output, and this leads to insufficient loosening of the heap.

References

[1] Eze V C *et al* 2017 *Chem Eng J* **322** 205–214 doi: 10.1016/j.cej.2017.04.038
[2] McDonough J R *et al* 2015 *Chem Eng J* **265** 110–121 doi: 10.1016/j.cej.2014.10.113
[3] Zhang J *et al* 2011 *Bioresour Technol* **102(16)** 7407–7414 doi: 10.1016/j.biortech.2011.05.035
[4] Mohayeji M *et al* 2016 *J Taiwan Inst Chem E* **60** 76–82
[5] Chen Z *et al* 2015 *Chem Eng Sci* **130** 254–263
[6] Borikov V N *et al* 2016 *J Phys Conf Ser* **671(1)** 012014 doi: 10.1088/1742-6596/671/1/012014
[7] Wilson B C *and* Patterson M S 2008 *Phys Med Biol* **53(9)** R61–109 doi: 10.1088/0031-9155/53/9/R01

IOP Conf. Series: Materials Science and Engineering **457** (2018) 012006 doi:10.1088/1757-899X/457/1/012006

Pressure determination of the water baffle shift in the capillary

I I Maksimov[1] and V V Alekseev[2]

[1]Professor, Mari State University, Yoshkar-Ola, Russia
[2]Associate Professor, Cheboksary Cooperative Institute of the Russian University of Cooperation, Cheboksary, Russia
E-mail: av77@list.ru

Abstract. The ability of soils to pass water characterizes water permeability, which is an important characteristic, depending on the type of soil, porosity and pore size distribution. It is estimated using the filtration coefficient k. To have information on the value of water permeability is necessary when creating irrigation systems, reservoirs, canals, dams and many other cases. Water permeability allows calculating water losses from these structures. The use of such technique as the creation of multi-layer thin-layer shields is effective to prevent the loss of water during its filtration from irrigating canals. Dissimilar materials, which leak in case of single laying-down, can dramatically decrease the loss of water being multiply used. The pore sizes in the soils vary greatly, which leads to the appearance of 'Jarman's chains' - alternating thin holes with thickenings. Water is trapped in the narrowest places of the capillary and is located in separate areas. Therefore, this kind of capillaries can be a gas-waterproof screen. The menisci layers formed at the contact, which exert considerable resistance to water filtration. Entrapped air bubbles are surrounded by water baffles. The article, according to the theory of the capillary potential (potential is considered as work on transferring a unit of mass of liquid from a capillary into a free liquid at the same temperature and at the same altitude level). The theory allow study the pressure that needs to be used to such a baffle to make it move.

1. Introduction

Efficiency and productivity of irritating systems are mostly defined by the loss of water during its filtration from irrigating channels. One of the techniques to prevent filtration is the creation of multi-layer thin-layer shields on the surface of the canal or the groundmass that forms the canal bed. Materials, which leak in case of single laying-down, being put layer by layer, form menisci that prevent the water filtration [1, 2]. Entrapped air bubbles, which cover the capillary, influence the movement of liquid in the capillaries. In most cases, these bubbles are divided by water baffles. This research is devoted to the determination, according to the theory of the capillary potential, of the pressure that needs to be used to such a baffle to make it move.

The height to which the liquid rises in a capillary is connected with its radius according to the equation of J. Jurin, which states that the thinner is a capillary the more the water would rise. Although examination of the distinct grounds shows that this equation is true only to a certain extent. If the diameter of capillaries is less than 10^{-8} m, Jurin's equation cannot be applied.

In such thin capillaries, there is no space for the capillary water because all the space is filled with the funicular and adsorbed water [3, 4].

That is why the height to which the liquid rises in the capillaries from the layer of the ground waters is bigger in clays than in heavy clays (see Table 1). Table 1 also includes the radii of the effective capillaries and the heights determined according to the equation of J. Jurin.

Moreover, if we simply take the soil pore space for the cylindrical one, the capillary in soils always has a variable diameter. In the interchange of thin holes with thickenings the water, in case of partial pore filling, will be only in the narrowest parts of the capillary in the form of the so-called 'Jarman's chains'. It is rater hard to take the water out of such a capillary.

Table 1. Height of the Capillary Water Rise (average readings)

Soil/Ground	Height of the capillary water rise, m	Radius of the effective capillary, 10^{-5}m
Sands	0.4–0.8	1.88–3.75
Sandy loams	0.8–1.2	1.25–1.88
Sandy clays	1.2–3.5	0.42–1.25
Clays	3.5–6.0	0.25–0.42
Heavy clays	up to 4.5	up to 0.33

If we apply pressure to one of the ends of a capillary, the menisci, which by their concavity face the pressure, will begin to sag more while the meniscus water holding forces will increase. The liquid is detained in such 'Jarman's chains'. That is why such capillaries can be gas- and waterproof shields [5–7].

In reclamation for water accumulation and tabulation to sandy soils of arid regions, in the construction of canals and reservoirs, the almost impermeable confining system is obtained by alternately stacking layers of dry sandy and loamy soils.

2. Experimental

Let's examine the capillary potential as a work of transferring of a liquid mass unit from the capillary to a free liquid at a constant temperature and at the same level of height. In this case, specific volume energy of the free liquid can make the zero level, which can be used to count the specific energy in the capillary.

Let's examine the system that consists of a dry capillary and some free liquid (Figure 1).

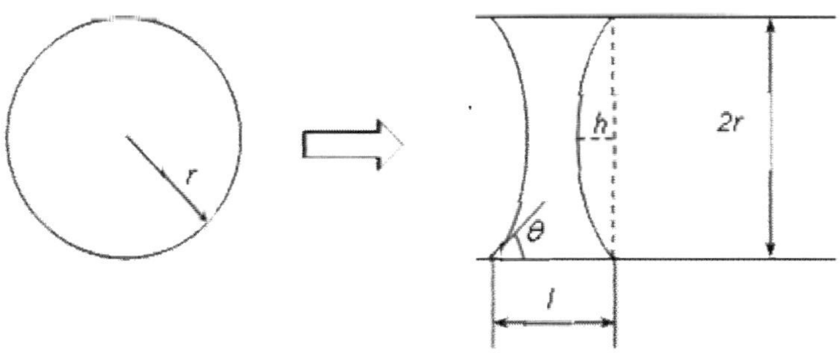

Figure 1. The Formation of the Baffle.

The total energy of such a system is defined by the surface energy of the capillary because the energy of the free liquid is equal to zero.

$$E_0=2\pi r l \sigma_{sg}, \quad (1)$$

where r is the radius of the capillary, l is the length of the capillary, and σ_{sg} is the free energy of the area unit of the dry capillary (i.e. solid – gas interfaces).

After the contact of the capillary with the liquid, the energy of the system can be expressed by the equation

$$E=2\pi rl\sigma_{sl} + 2\pi(r^2+h^2)\,\sigma_{lg} + A, \qquad (2)$$

where $2\pi rl\sigma_{sl}$ is the surface energy of the solid – gas interface, $2\pi(r^2+h^2)\sigma_{lg}$ is the surface energy of the liquid – gas interface (h is height of the spherical segment), and A is a load of the introsusception of the liquid into the capillary, equal to $\pi rl\sigma_{lg}$.

As the liquid is not free anymore, the "decrease" of its energy during the introsusception into the capillary is defined by the difference of E_0 - E, i.e. the equation:

$$\Delta E=2\pi rl\sigma_{sg} - 2\pi rl\sigma_{sl} - 2\pi(r^2+h^2)\,\sigma_{lg} - \pi rl\sigma_{gl}\,, \qquad (3)$$

or, if applying the famous equation of Neiman,

$$\Delta E=2\pi rl\sigma_{gl} - 2\pi(r^2+h^2)\,\sigma_{lg} - \pi rl\sigma_{gl}\,. \qquad (4)$$

For the convenience of calculations, we will consider such a free liquid that is equal in volume to the volume of a sphere, whose radius is equal to the radius of the capillary. In the case of an incompressible liquid, taking into consideration the immutability of the volume of liquid in the process of being drawn into the capillary, we can write:

$$\frac{3}{4}\pi r^3 = \pi r^2 l - \frac{2}{6}\pi h(3r^2 + h^2)\,. \qquad (5)$$

Now let's define the length of the wetted part of the capillary through the height of the spherical segment h.

$$l = \frac{4r^3 + 3r^2 h + h^3}{3r^2}\,. \qquad (6)$$

Let's apply l from the equation (6) to the equation (4) and take the derivative $\frac{d(\Delta E)}{dh}$. While equating this derivative to zero, we will find out h that determines the maximum of the system's energy loss, i.e. the system remains in the condition of the energy development:

$$h = -\frac{4r \pm \sqrt{16r^2 - 4r^2}}{2r}\,. \qquad (7)$$

Taking into consideration that $h \le l$, we get:

$$h = (2-\sqrt{3})r = Cr\,, \qquad (8)$$

where C is the derivative.

All the expressed ideas are based on the theory that a liquid can wet a capillary rather substantially, i.e. the interface angle of wetting θ while its general determination is equal to zero.

In real conditions $\theta = 0$ only for very clean and non-adsorbed envelopes. Usually, the interface angle is not equal to zero. That is why to determine h you need to consider the angle θ.

From the simple geometrical relationships, we can show that $h=Br$, where

$$B = \frac{1-\sqrt{1-\cos^2\theta}}{\cos\theta}\,. \qquad (9)$$

Generalizing the equations (8) and (9), we can eventually get

$$h=CBr. \qquad (10)$$

From the geometrical relationships, it is clear that:

$$\cos\theta_1 = \frac{2h}{1+C^2B^2}, \tag{11}$$

That allows finding out the interface angle θ of a baffle in case of its full wetting. For the water in a glass capillary $\theta_1 = 60^\circ$.

Let's find the capillary potential for a baffle

$$\varphi = \frac{\Delta E}{m} = \frac{\Delta E}{V\rho}, \tag{12}$$

where m is the mass of the liquid; V is the volume of the liquid; ρ is liquid density.

Let's apply the theory of dimensions.

$$\frac{\text{J}}{\text{kg}} = \frac{\text{N}\cdot\text{m}}{\text{kg}} = \frac{\frac{\text{N}}{\text{m}^2}}{\frac{\text{kg}}{\text{m}^3}} \tag{13}$$

thus,

$$\varphi = \frac{\Delta p}{\rho}.$$

Equating (12) to (13), we will get

$$\frac{\Delta E}{V\rho} = \frac{\Delta P}{\rho} \quad \text{or} \quad \Delta P = \frac{\Delta E}{V} \tag{14}$$

where ΔP is the pressure change in the liquid in the capillary relative to the pressure in the free liquid at the same altitude level. From the equation (4) and (14) we can derive that:

$$\Delta P = \frac{2\pi r l \sigma_{gl}}{V} - \frac{2\pi(r^2+h^2)\sigma_{gl}}{V} - \frac{\pi r l \sigma_{gl}}{V} = \Delta P_1 - \Delta P_2 - \Delta P_3 \quad , \tag{15}$$

where ΔP_1 is the internal pressure created by forces of interaction of liquid molecules with a solid wall that does not change, at least until the beginning of the movement of the baffle; ΔP_2 is internal negative pressure, the "tension" of the liquid generated by the liquid-gas interface, which can be assumed unchanged up to the beginning of the movement; ΔP_3 is internal negative pressure associated with the retraction of the liquid into the capillary, which may change if you change the shape of the baffle under the unilateral external pressure.

The condition of the begging of the baffle's movement – its extreme points – is Equation

$$|\Delta P_3| = \Delta P', \tag{16}$$

where $\Delta P'$ is external gas pressure on the baffle, on the one hand, under which the extreme points of the baffle start to move or "shear" pressure.

Thus, the required "shear" pressure

$$\Delta P' = \frac{\pi r l \sigma_{gl}}{V}, \tag{17}$$

where $V = \pi r^2 l - \frac{2}{6}\pi h(3r^2 + h^2)$ is the amount of liquid in the baffle.

Taking h from (10) to the equation (17), after the transformations we will get

$$\Delta P' = \frac{3l}{3l - r\,(3CB + C^3B^3)}\,\frac{\sigma_{gl}}{r} \quad (18)$$

3. Results and Considerations

Table 2 shows the results of calculations of shear pressure at the effective radius of the capillaries in the soil that are given in Table 1. As can be seen from Table 2, for the used in hydraulic engineering and reclamation alternately stacking of a couple of layers of dry sandy and loamy soils the difference in pressure is in the range from 2 to 15.5 kPa, i.e. even one interface between the layers can keep the pressure of the water column from 0.2 to 1.5 m.

Table 2. Shear Pressures (average readings)

Soil/Ground	Shear pressure, Pa
Sands	1947–3883
Sandy loams	3883–5840
Sandy clays	5840–17381
Clays	17381–29200
Heavy clays	22121

4. Summary

There was proposed and tested the equation which allows determining the pressure that must be applied to the liquid baffle to move it through the capillary. The results can be used in hydraulic engineering and land reclamation in the design of the objects of water accumulation and inlet, because depending on the technical parameters of the object you can pick up both types of soil for stacking of layers and the required number of layers. Filtration losses, in this case, are minimized and the calculation of multi-layer thin-layer shields gets a theoretical basis.

References

[1] Eze V C *et al* 2017 *Chem Eng J* **322** 205–214
[2] McDonough J R *et al* 2015 *Chem Eng J* **265** 110-121 doi: 10.1016/j.cej.2014.10.113
[3] Alekseev V V *and* Maksimov I I 2013 *Eurasian J. Soil Sci* **46(7)** 751–757 doi: 10.1134/S1064229313070028
[4] Sysuev V A *et al* 2013 *Russian Agricultural Sciences* **39(5-6)** 522–525 (in Russian)
[5] Chen Z *et al* 2015 *Chem Eng Sci* **130** 254–263
[6] Hodgkinson J *and* Tatam R P 2013 *Meas Sci Technol* **24** 012004 doi: 10.1088/0957-0233/24/1/012004
[7] Wilson B C *and* Patterson M S 2008 *Phys Med Biol* **53(9)** R61–R109 doi: 10.1088/0031-9155/53/9/R01

IOP Conf. Series: Materials Science and Engineering **457** (2018) 012007 doi:10.1088/1757-899X/457/1/012007

Formation and impact of soil particles on flame radiation in the cylinder of diesel while working on gas-motor fuel

V A Likhanov[1] and A V Rossokhin[2]

[1]Professor, Vyatka State Agricultural Academy, Kirov, Russia
[2]Associate Professor, Vyatka State Agricultural Academy, Kirov, Russia
E-mail: rossokhin.dvs@mail.ru

Abstract. The article deals with the formation of soot particles in the cylinder of a high-speed diesel engine with turbocharging when operating on gas engine fuel (compressed natural gas). It is shown that the main source of data on the processes occurring in the engine cylinder is the indicator diagram. But the data obtained with its help is often not enough to answer a number of questions regarding the characteristics of the course of intra-cylinder processes. As an example, one can cite the problem of studying the formation of harmful substances in an engine cylinder during fuel combustion, such as soot (C), the mechanism of the chemical kinetics of the formation of which has not yet been explicitly solved. The article presents the spectral effective cross-sections of the absorption of soot particles, depending on the optical properties and dispersion, spectral and integral absorption coefficients of soot particles, depending on the angle of rotation of the crankshaft.

1. Introduction

The intensity of combustion of fuel, which determines the nature and dynamics of heat release, depends on the physical and chemical processes occurring in the reaction zone. And the conditions in which these reactions occur, especially in engines with internal mixture formation, undoubtedly have a pronounced dependence on the location of the local volume of the combustion chamber in question.

Another interesting, important, but little studied process, is the process of radiative heat transfer in the combustion chamber of a diesel engine. The complexity of studying this phenomenon is due to the presence in the combustion chamber of the diesel engine of a large number of soot particles of different shapes and sizes, whose emissivity significantly exceeds the emissivity of gases, and the processes of formation and, accordingly, burnout of these particles have a clearly expressed local character. And to calculate the processes of formation and burning out of soot, relying solely on the indicator diagram can only be in the first approximation.

For the combustion chambers of diesel engines in calculating radiation heat transfer, it is necessary to take into account a list of the features of the working process associated with the cyclicity and instability of the working process, and with the geometrical parameters of the combustion chamber and the fuel flare. Depending on the intensity of the turbulence of the air charge in the combustion chamber and the parameters of the top jet spray injected by the injector, a volume of gases containing soot particles is formed. The presence of a certain duration of fuel supply, the different sizes of droplets in the torch, the set of accidental spontaneous ignition points, the turbulence of the air charge in the cylinder as a result of the movement of the piston and burning of the fuel, the varying volume of the cylinder, the continuously changing amount of soot particles and their dimensions practically reduce zero possibility of direct calculation of the amount of soot particles in the local volume and their dimensions, which is required for calculations of radiative heat transfer in the combustion chamber.

The processes of injection and spraying of fuel have been studied in sufficient detail. Knowing the fuel supply law, the injection pressure, the diameter of the spray holes, the physical and chemical properties of the fuel, the conditions in the combustion chamber can be determined with sufficient accuracy for further calculations to determine the size, shape and process of the fuel-air jet development. It is common knowledge that the main part of the combustion and heat release process is completed in 50...60o r.c. after TDC. It is during this period that the main amount of fuel burns in the volume of the combustion chamber when the piston is near TDC, the pressure and temperature of the gases in the engine cylinder reach the maximum values, as well as the concentration of soot particles. This, in turn, leads to the fact that the emissivity of the flame is high.

2. Experimental

A feature of the processes of formation and burnup of soot particles in the cylinder of the engine with internal mixture formation is a significant heterogeneity in the composition of the fuel-air mixture in the combustion chamber and its large temperature unevenness. That is, the processes of soot formation and further oxidation of solid particles are of a local or band nature and depend on a number of factors, although the stages or phases of these processes can be formulated as follows.

Phase I is the phase of predominantly soot formation. The beginning of the formation of soot particles coincides with the onset of visible combustion and the active release of heat in the engine cylinder during the combustion of the fuel supplied for the ELV.

Phase II is the phase of simultaneous formation and oxidation of particles, lasting until reaching the greatest saturation. This phase corresponds to a second maximum of the soot emission rate, whose position lags behind the second maximum of the heat release rate. At this moment, the diffusion combustion of the main portion of the fuel takes place in the cylinder. The combustion process takes up most of the combustion chamber, and the temperature of the working fluid reaches its maximum values. The processes of formation and oxidation of soot particles come at comparable rates, but the quantitative mass yield of soot exceeds the amount of burnt soot, and the rate of the final process remains positive, causing an increase in the number of soot particles to their peak values.

Phase III is the phase of preferential oxidation of soot particles, which lasts until the opening of the exhaust valves. This phase corresponds to a third maximum of the rate of soot emission, which assumes negative values, since it characterizes the reduction of the containment in the cylinder. The process of combustion by this time has ended, and only the burning of individual sections of unburned soot occurs. The formation of soot is also almost complete, and the burning out of the remaining soot from the second phase continues. The more time allowed for the burning of soot, the less it remains in the cylinder at the time of release [1].

The quality of spraying of fuel depends on its viscosity, pressure in the compressor, geometric parameters of the atomizer, fuel delivery intensity, etc. At the same time in the volume of the torch there will always be a drop of fuel, differing in both size and fractional composition. And since the dynamics of combustion and soot formation is determined mainly by these parameters, and also by the intensity of the turbulence of the air flow in the CS, reliable data for calculating radiative heat transfer can be obtained only on the basis of their experimental determination.

According to the data given in [2], the burning rate of soot can be estimated, for example, from the fact that the initially formed particles have a certain size, and at the end of the working stroke they completely burn out. The particle size found in combustion products in diesel engines can be estimated at $(0.2...1.0) \times 10^{-6}$ m. For an engine with a crankshaft rotation speed of 1500 min^{-1}, the visible burning time is $(7...8)\times10^{-3}$ s, i.e. at a linear rate of reduction of the particle radius, the burning rate of the soot can be estimated at $2.5\times10^{-5}...1.25\times10^{-4}$ m/s. If the density of carbon black is 2×10^{3} kg/m^3, then the average heat release rate on the surface of the soot is 850...4300 W/m^3. Part of this energy goes to the heating of combustion products, some are transferred to the environment by convection, some by radiation. The approximate value of the heat capacity of the reaction products is estimated as the sum of the specific heats of the components, for example, CO_2 and N_2. With the burn-up rate adopted above, the heat capacity of the combustion products produced per unit time from a unit of the soot surface is 16.2×10^3 W/(m$^2\times$K). In order to estimate the intensity of convective heat transfer for the adopted conditions, the heat transfer coefficient α will be $2.5\times10^4...1.25\times10^5$ W/(m$^2\times$K).

Thus, with the accepted burn-up rate of the soot particles, only due to the mechanism of conductive heat exchange, all energy released is discharged into the environment when the temperature of the soot particle of the ambient temperature exceeds the temperature by less than 1 K. This allows us to further consider the combustion products, containing soot particles, as a homogeneous radiating, absorbing and scattering medium [3].

At the same time, soot particles are located not only in the volume of the cylinder, but also in the so-called boundary layer, i.e. layer adjacent to the walls of the cylinder and CC. The thickness of this layer is 2...3 mm at the end of the expansion cycle. It was noted in [4] that near the wall, soot particles have a lower temperature than in the core of the torch. Under these conditions, combustion of the soot particles is slowed down, so they can not be any appreciable sources of heat release. On the basis of this, it is concluded that the intensity of convective heat transfer in the cylinder is determined by the flow conditions and the state of the gas in the immediate vicinity of the wall, while the intensity of radiant heat transfer depends on the concentration and temperature of the combustion products, including soot particles, in the volume of the cylinder .

To determine the intensity of radiative heat transfer in a medium containing suspended particles, in addition to the boundary conditions, the spectral absorption and scattering coefficients of the particles and the scattering indicatrix of the radiation on the particles must also be known. These characteristics, in turn, depend on the physical nature of the substance of the particles (the complex refractive index of the particle material), their dimensions, concentration, shape and wavelength of radiation. The most famous papers devoted to the solution of the problem of finding scattering and absorption indices on various particles include works [5].

Thus, in [5] the distribution of soot particles in size is given under different conditions (Figure 1). In this case, the distribution can be normal (curves 1 and 2), close to normal and asymmetric (curve 3).

In the same paper, an analysis is given of the calculated results for the concentration of soot particles Z as a function of particle size and mass fraction (Figure 2).

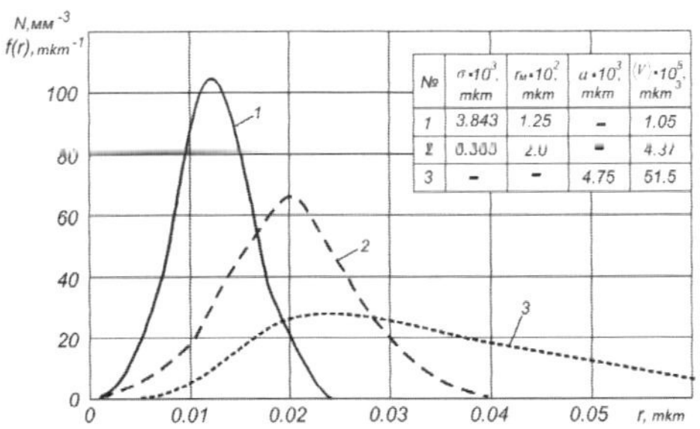

Figure 1. Distribution of soot particles by size.

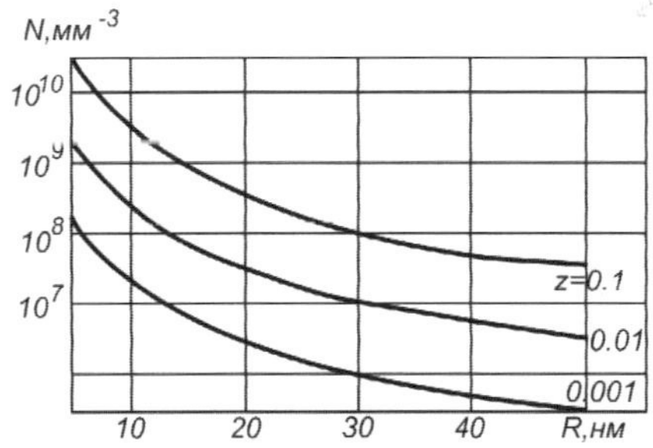

Figure 2. Concentrations of soot particles Z depending on the particle size and mass fraction.

An analysis of the results shows that a decrease in particle size leads to a sharp increase in their concentration.

Using these and other results, it is possible to determine the scattering and absorption indices of the emission of a cloud of soot particles. At the same time, analytical solutions are obtained only for particles of simple and regular geometric shapes, for example spheres.

In practical calculations of heat transfer by radiation in the combustion engine, the characteristics of the medium are usually used on the average along the length of the beam, while the emitting capacity of the torch is calculated from $a_v = 1 - \exp\left(\frac{N}{V}\frac{\pi d^2}{4}\vartheta_v \alpha\right)$, where d is the effective particle diameter. It should be borne in mind that to determine the emission characteristics of soot particles, reliable data are required as to the concentration and distribution of soot particles in size, and their complex refractive index.

In this case, the value of the complex refractive index varies with the temperature change, and this dependence is not exactly determined. The diameter of the particles also changes with time, and the H / C ratio also changes, and this in turn affects the optical properties of soot particles.

At the same time, there are no sufficiently accurate methods for calculating radiative heat transfer in fast-moving autotractor diesels due to the extremely short reaction time and a very large number of factors affecting it.

To determine the radiative characteristics of the sooty flame, it is necessary to know the optical properties of the solid particles on which the refractive index of the medium depends:

$$m = n_1 - n_2 \cdot i. \tag{1}$$

The rate of propagation of radiation in a gaseous medium depends on this index. The absorption index of the medium n_2 determines the decrease in the amplitude of the electromagnetic oscillations caused by the absorption in the medium. These optical constants n_1 and n_2 depend on a number of parameters of the substance in which the radiation is propagated, such as the temperature of the substance, the wavelength of the radiation, the presence of suspended particles or a homogeneous gas medium, etc.

The optical constants n_1 and n_2 for carbon blacks typical for combustion chambers of diesel engines, depending on the temperature and the ratio of H/C for different wavelengths are shown in Figure 1.

If we consider a single carbon black particle, which has the form of a ball of a certain radius r, through which the radiation wave passes, then part of the energy of this electromagnetic wave will be absorbed by the particle, and some will be scattered. When studying the indices of individual particles, the following parameters are used: the effective cross sections for attenuation, scattering and absorption are determined by the Equations:

$$\sigma_{att} = \pi r^2 K_{att}(m,\rho), \quad \sigma_{sc} = \pi r^2 K_{sc}(m,\rho), \qquad \sigma_{abs} = \sigma_{att} - \sigma_{sc}, \tag{2}$$

where K_{att} and K_{sc} are the dimensionless attenuation and scattering efficiency factors, which depend on the parameters m and $\rho = 2\pi r / \lambda$.

The values of K_{att} and K_{sc} are written in terms of the relations:

$$K_{att} = \frac{4}{\rho^2}\sum_{n=1}^{\infty}(2n+1)\mathrm{Re}(a_n + b_n), \qquad K_{abs}(m,\rho) = \frac{2}{\rho^2}\sum_{n=1}^{\infty}(2n+1)\left(|a_n|^2 + |b_n|^2\right), \tag{3}$$

where a_n and b_n are the amplitudes of the n-th electric and n-th magnetic waves, respectively (the Mi coefficients).

When electromagnetic radiation is propagated in a substance that is solid particles of the same particle size and composition in the amount of N in a single volume suspended in the gas, it is necessary to consider the characteristics of a single volume rather than an individual particle. It is characterized by spectral coefficients of attenuation, scattering and absorption:

$$\kappa_\lambda = N \cdot \sigma_{att}, \qquad \beta_\lambda = N \cdot \sigma_{sc}, \qquad \alpha_\lambda = N \cdot \sigma_{abs}. \tag{4}$$

In this case, the value of N can be expressed in terms of the mass concentration of soot particles C_m:

$$N = \frac{3C_m}{4\pi\rho r^3}, \tag{5}$$

where ρ is the particle density (for carbon black $\rho = 1.9$ g/cm^3).

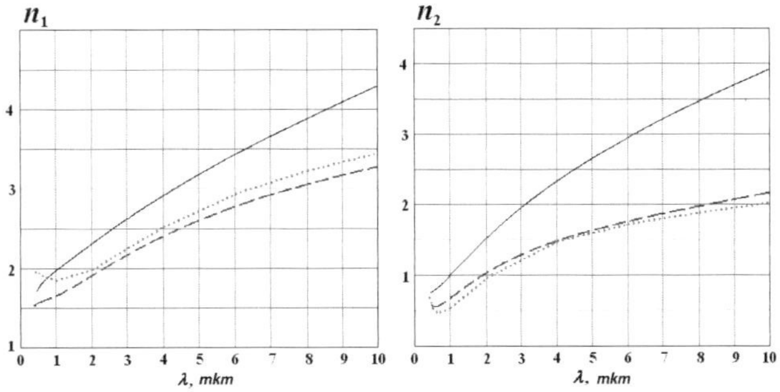

Figure 3. Dependence of optical constants and black carbon on the wavelength, where ——— at H/C=0, T = 2250 K [6]; •••• at H/C=0.07…0.22, T = 1700 K [7], — — at H/C=0.217, T = 293 K [6].

For conditions corresponding to the conditions for the formation of soot particles in the engine cylinder, the presence of particles of different sizes is characteristic, that is, it is no longer a monodisperse but a polydisperse medium. In this case, the following dependencies should be considered:

$$\sigma_{att} = \int_0^\infty \pi r^2 K_{att}(r) \cdot f(r) \cdot dr\,, \quad \sigma_{sc} = \int_0^\infty \pi r^2 K_{sc}(r) \cdot f(r) \cdot dr\,, \qquad \sigma_{abs} = \sigma_{att} - \sigma_{sc}\,. \tag{6}$$

$$\kappa_\lambda = N \cdot \int_0^\infty \pi r^2 K_{att}(r) \cdot f(r) \cdot dr, \; \beta_\lambda = N \cdot \int_0^\infty \pi r^2 K_{sc}(r) \cdot f(r) \cdot dr, \; \alpha_\lambda = N \cdot \int_0^\infty \pi r^2 K_{abs}(r) \cdot f(r) \cdot dr. \tag{7}$$

Then the value of N is defined as:

$$N = \frac{3C_m}{4\pi\rho \int_0^\infty r^3 f(r) dr}\,. \tag{8}$$

3. Results of Researches

For the practical application of this technique for studying the radiative characteristics of matter, we studied the working process of a high-speed diesel engine with a turbocharger of the dimension 4FP 11.0/12.5 when operating both on diesel fuel and on gas engine fuel. As a calculation, we considered the nominal operating mode of the diesel engine.

The distribution of soot particles in the exhaust gases in size is described by the distribution function:

$$f(r) = \frac{4r^2}{\sqrt{\pi} r_m^3} e^{-\left(\frac{r}{r_m}\right)^2}\,. \tag{9}$$

We established the dependence of the radiative properties of soot particles on their optical parameters and the particle size. The results of the calculation of the characteristics of individual soot particles are presented using data on optical constants for both particle systems of the same size and for polydisperse systems in the wavelength range 0.5...10 mm.

Further, the dependence of the radiation characteristics of the final volume on the angle of rotation of the crankshaft during the operation of the diesel engine on diesel fuel and compressed natural gas was investigated. Table 1 shows the change in the mass concentration of soot particles in the diesel and gas diesel cylinder at different angles of rotation of the crankshaft (s.q.v.).

Table 1. Dependence of the mass concentration of soot in the cylinder of a diesel engine and gas diesel engine of the dimension 4FP 11.0/12.5, depending on the change in the angle of the pc.

	s.q.v. φ, [°]							
	5	10	15	20	40	60	80	100
Mass concentration C_m, g/m^3 (diesel)	0.15	0.32	0.33	0.32	0.27	0.23	0.18	0.15
Mass concentration C_m, g/m^3 (gas diesel engine)	0.10	0.20	0.23	0.22	0.18	0.15	0.11	0.08

For a system of soot particles characterized by the presence of particles of different sizes, the largest value of the spectral (α_λ) and integral (α) absorption coefficients is observed at an angle φ = 15°s.q.v. (Figure 4*a*). The values increase with decreasing wavelength from 10 mkm to 0.5 mkm. For a gas-diesel engine, the dependence for is the same, although the numerical values are reduced by 30...47% (Figure 4*b*).

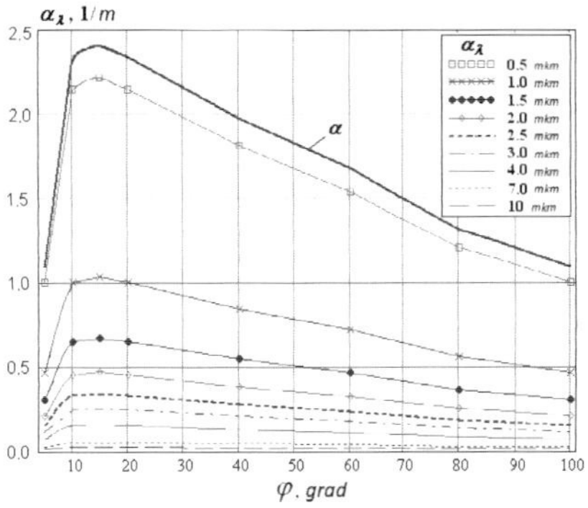

a

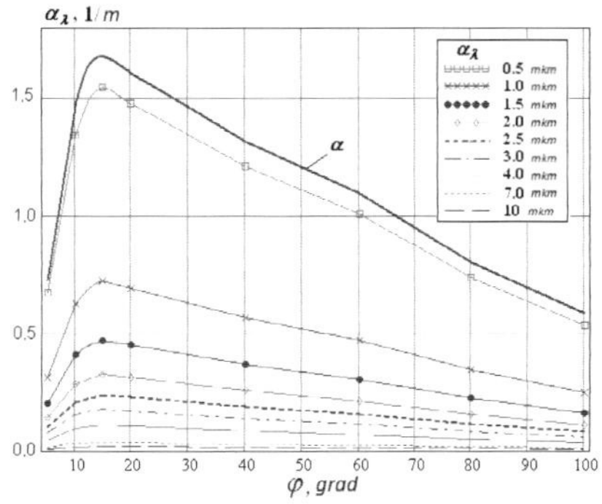

b

Figure 4. Spectral and integral absorption coefficients of soot particles as a function of the angle of the pcc, where a is diesel process; b is gas-diesel process.

4. Conclusion

The results of the calculations made it possible to determine the radiation characteristics of soot particles, depending on their optical properties and dimensions. The calculated values of the radiation characteristics of soot black by optical constants at T = 293...2250 K lead to a discrepancy of the results up to 45%, which indicates the influence of the temperature of the medium on these processes.

The use of a monodisperse system in the calculations instead of a polydisperse system leads to a difference of up to 8% in the presence of PXEO. At the RHICH level, with a similar replacement, the results may differ by more than 2 times. The conducted comparison of PXEO with respect to the diesel engine and gas diesel engine 4FP 11.0/12.5 at different angles of rotation of the crankshaft is necessary for further estimation of the share of heat exchange by radiation when switching to gas engine fuel. This will allow to determine the working and technical characteristics of diesel engines working on alternative fuels.

References

[1] Likhanov V A *and* Lopatin O P 2017 *Thermal Engineering* **64(12)** 935–944
[2] Klein M A 2004 *Specific Heat Ratio Model and Compression Ratio Estimation* (UniTryck: Sweden)
[3] Pickett L M *and* Siebers D L 2004 *SAE International Journal of Fuels and Lubricants* **113(4)** 614–630
[4] Abbaszadehmosayebi G *and* Ganippa L 2014 *Applied Energy* **122** 143–150
[5] Ghojel J *and* Honnery D 2005 *Applied Thermal Engineering* **25(14-15)** 2072–2085
[6] Rakopoulos D C *et al* 2011 *Fuel* **90** 1855–1867
[7] Abbaszadehmosayebi G *and* Ganippa L C 2014 *Fuel* **119** 301–307

IOP Conf. Series: Materials Science and Engineering **457** (2018) 012008 doi:10.1088/1757-899X/457/1/012008

The Accounting of heat losses and infiltration during the work with a microclimate of rooms in broiler production

A I Matveev[1], A M Rybakov[1], A F Nasibullina[1], I V Gantimurova[1], A L Rozhentsov[2]

[1]Master Degree Student, Mari State University, Yoshkar-Ola, Russia
[2]Associate Professor, Mari State University, Yoshkar-Ola, Russia
E-mail: energetik-84@mail.ru

Abstract. The condition of the air environment in modern agricultural buildings is provided by means of the microclimate air conditioning systems (MACS). However traditional MACS and their equipment are quite expensive, power-intensive and difficult in operation. Therefore relevant is a problem of development and use of simpler, reliable, energy-efficient systems and devices of conditioning of a microclimate for creation and maintenance of the required technological parameters of the air environment providing decrease in energy consumption and increase in efficiency in poultry farming has important social value. Air conditioning system of a microclimate of the agricultural building has to represent complex application active (air conditioning systems) and passive (external protections) engineering means. Need of recirculation of air is caused by considerable overexpenditures of warmth during the cold period of year and cold during the warm period of year. It is connected with the fact that norms of air exchange for the warm and cold periods of year differ several times, at the same time the total value of air supply remains to a constant. Infiltration is an unorganized air intake to room through leakage of buildings protections under the influence of thermal and wind pressure and, possibly, due to the work of mechanical ventilation. Accounting losses of heat in poultry-farming rooms and dispensing of ventilation volume, the major factor is the air infiltration through construction and enclosing structures that significantly affects the gas consumption for space heating and on the final cost of production.

1. Introduction

Currently there is an active reconstruction of premises for maintenance of animals and other objects for premises for maintenance broiler chickens. Since these premises were built many years ago, heat-insulating and air-tightness of such buildings are poor [1–3].

Broiler chickens are very demanding on the parameters of microclimate. Calculating the minimum ventilation, the broiler production manager must know the amount of infiltration losses which should be considered when determining the minimum air exchange every day. In addition, it is necessary to know how much heat is lost through the enclosing structures. In the case of large heat losses through the enclosing structures, it is possible to offer a feasibility study on insulation of premises with a payback period of investment [4–6].

2. Choice of Calculation Method

There are two methods of calculating losses through enclosing structures and infiltration:

1. by using the gas flow spent on compensation for losses (in the case of a gas flow meter).

2. by the time of operation of gas generators (in the absence of gas flow meter).

Since many poultry-farming premises do not have a gas flow meter (usually a single meter for several premises), the second option is more preferable. Nevertheless, the option of determining losses through enclosing structures and infiltration by using the gas flow spent on compensation for losses is the most accurate and requires detailed consideration.

The determination of losses through enclosing structures and infiltration, regardless of the selected method, is carried out in 3 stages: the determination of heat losses through enclosing structures with disconnected combined extract-and-input ventilation, the determination of total losses for infiltration and through enclosing structures with included exhaust ventilation, and the determination of losses amount for infiltration.

Let consider in more detail the steps in the case of choosing the definition of heat loss and infiltration by using the time of gas generators operation.

3. Results and considerations

Since many poultry-farming premises do not have a gas flow meter (usually a single meter for several premises), the second option is preferable. The determination of losses through enclosing structures and infiltration is carried out in 3 stages.

3.1 The determination of heat losses through enclosing structures with turned-off ventilation

3.1.1 Phase 1

The result will be most accurate if measurements are taken in the dark or in cloudy weather [7].

It is necessary to turn off exhaust ventilation, to close all extract and input valves, to pack off and to insulate tunnel exhaust fans and extract tunnel louver, all doors should be closed. Then, install the gas generators in the automatic position. During at least 24 hours, the poultry house must be heated to a temperature of 32.5^{0}C. At the same time the temperature of the house should be higher than the temperature of the street.

After warming up the poultry house, it is possible to begin to carry out the corresponding measurements of heating operating time. Modern microclimate controllers automatically record the heating time. During the measurements it is necessary to make sure that all the heaters are in good order and the settings correspond to the declared power.

The measurements should be at least 3 with an interval of 60 minutes. At the time of each measurement it is necessary to record outdoor and indoor temperature.

Settlement part is below.

1. The general operating time on each group t_k, for all period of measurements, minute, min

$$\mathrm{t_k} = \sum_{k=1}^{n} t_{ki} , \qquad (1)$$

where t_{ki} is operating time of i measurement of *(k)* group, min; *n* are measurements; *k* are groups.

2. Pays off heat from *(k)* group of heating N_k, kW.

$$N_k = \frac{P_k \times t_k \times L_k}{60} , \qquad (2)$$

where P_k is power of a gas generator in the *(k)* group, kW; L_k is number of gas generators *(k)* group, pcs.

3. Total heat from all groups *(N)*, kW

$$N = \sum_{k=1}^{m} N_k , \qquad (3)$$

IOP Conf. Series: Materials Science and Engineering **457** (2018) 012008 doi:10.1088/1757-899X/457/1/012008

where m are groups.

4. The difference between internal and external temperature T, ^{0}C

$$T = T^{building} - T^{outdoorair} \tag{4}$$

5. Heat losses through enclosing structures are determined q_{losses}, kW/^{0}C

$$Qpoten. = \frac{N}{(n \times T)} \tag{5}$$

3.1.2 Phase 2. The determination of heat losses for infiltration with turned-on ventilation

The result will be most accurate if measurements are taken in the dark or in cloudy weather. This stage is a continuation of the 1st stage and all its conditions must be met except that the ventilation will be tuned to the working static division (all inflow is closed at the same time). At the same time air will pass through defects of building structures and leakage of inlet valves. The measurements should be at least 3 with an interval of 60 minutes. At the time of each measurement it is necessary to record outdoor and indoor temperature.

1. The required quantities are determined by Equations (1), (2), (4).

2. Further it is necessary to find total heat from all groups spent for losses through enclosing structures and infiltration $N_{los,\ inf}$, kW

$$Npoten.\text{infil.} = \sum_{k=1}^{m} Nk \tag{6}$$

3.1.3 Phase 3. The determination of air volume for infiltration

Knowing the difference of the spent heat between the 2nd and 1st measurement steps, it is possible to calculate the volume of infiltration air. Infiltration will be the same during the whole period of the poultry, and this must be taken into account when calculating the minimum daily ventilation.

1. Amount of heat for infiltration, kW / h:

$$N_{\text{inf}.} = \frac{N_{poten.\text{inf}.} - N}{n} \tag{7}$$

2. Amount of infiltration, m^3/h.

$$V = \frac{360000}{C \times P_{outdoorair} \times T}, \tag{8}$$

where C = 1005 kJ/kg ^{0}C, air specific heat, $P_{outdoorair}$–solidity of outdoor air, kg/m^3, according to an Equation (9):

$$P_{outdoorair} = \frac{353}{273 + T_{outdoorair}} \tag{9}$$

3.2 The determination of heat loss with gas flow

3.2.1 Phase 1. The determination of heat losses through enclosing structures with disconnected extract-and-input ventilation

The conditions of experiment are the same as in the method considered above.

The record of readings of gas flow is necessary at the beginning and at the end of experiment. The readings record of indoor and outdoor temperature is necessary at the beginning and at the end of the experiment. The duration of the experiment is at least 3 hours.

Further sequence is below:

1. The determination of the amount of heat expended on losses through enclosing structures according to the Equation (10), kW:

$$N = 10.1 \times L, \tag{10}$$

where 10.1 is the lowest heat of gas combustion from Russia; L is the difference of counter reading, m^3.

2. The difference of indoor and outdoor temperature is calculated using, T ^0C the following Equation (4).

3. The heat losses through enclosing structures is calculated using q_{los}, kW/^0C the following Equation (5).

3.2.2 Phase 2. The determination of heat losses for infiltration with turned-on ventilation

The conditions of experiment are the same as in the method considered above.

The record of gas flow readings is necessary at the beginning and at the end of experiment. The readings record of indoor and outdoor temperature is necessary at the beginning and at the end of experiment. The duration of experiment is at least 3 hours.

Knowing the difference of the gas meter readings according to the Equation (11), we determine the amount of total losses through enclosing structures and infiltration, kW:

$$N_{los.\inf.} = 10.1 \times L_{los.\inf.}, \tag{11}$$

where $L_{los,\ inf}$ is the difference of the gas meter readings of the 2nd stage, m^3.

3.2.3 Phase 3. The determination of air volume for infiltration

Using Equations (7–9) we determine the amount of air passing through the leakage of structural elements of premises.

Therefore, working with the ventilation system every day it is important to consider all factors affecting on ventilation. At the same time infiltration is hardly the most significant size which influences on operation of ventilation system. It allows to dose with high precision air exchange in poultry-farming premises.

The determination of heat loss amount for infiltration and heat loss through enclosing structures must be made with a certain periodicity. It is connected with the fact that the majority of heat-insulating materials as well as joint sealing means have limited service life. It is recommended to make schedules of change of above-mentioned losses and, in the case of jumps, look for causes using devices which allow to define locally places in premises with the increased heat conductivity or depressurizations of walls joints, side valves or in other places.

4. Summary

Therefore, when you daily work with the ventilation system it is important to consider all factors affecting on ventilation. At the same time infiltration is hardly the most significant size which influences on operation of the ventilation system. It allows to dose with high precision air exchange in poultry-farming premises.

References

[1] Ghaderi M *et al* 2017 *Measurement* **114** 191–194
[2] Gutierrez M F *et al* 2017 *Chemosphere* **171** 644–653
[3] Alsobayel A A *et al* 2016 *J of the Saudi Society of Agricultural Sciences* doi: 10.1016/j.jssas.2016.03.001
[4] Ramli M B *et al* 2015 *Procedia Manuf* **2** 527–531
[5] Franco S C *et al* 2017 *Aquaculture* **468** 569–578
[6] Jura M *et al* 2000 *Astrophys J* **528** L105–108
[7] Minami K *et al* 2010 *J Phys Conf Ser* **215(1)** 012164 doi: 10.1088/1742-6596/215/1/012164

IOP Conf. Series: Materials Science and Engineering **457** (2018) 012009 doi:10.1088/1757-899X/457/1/012009

Research of a diametrical fan with suction channel

V E Saitov[1], V G Farafonov[2], R G Gataullin[3], A V Saitov[4]

[1]Professor, Viatka State Agricultural Academy, Kirov, Russia
[2]Associate Professor, Viatka State Agricultural Academy, Kirov, Russia
[3]Engineer, Viatka State Agricultural Academy, Kirov, Russia
[4]Student, Viatka State Agricultural Academy, Kirov, Russia
E-mail: vicsait-valita@e-kirov.ru

Abstract. Various grain cleaning machines equipped with pneumatic systems are widely used for cleaning grain from impurities, in which as an air-flow generator, diametric fans are increasingly used to create a uniform airflow over the width of pneumatic systems for high-quality grain cleaning. Installation of fans of this type in the pneumatic system of grain cleaning machines in some cases entails the creation of guiding elements in front of its intake part. A diametric fan with a suction channel located in the initial part of the spiral body and allowing it to be assembled into pneumatic systems of grain cleaning machines without additional or special taps has been developed. Theoretical and practical studies have been carried out to determine the depth Δh of the suction channel. It is established that the most rational value of the depth Δh of the suction channel is 0.05 m for a fan with an outer diameter D_2 of the impeller 0.30 m. The fan has three operating modes. With the suction canal open, suction of dusty air from the sludge chambers, dust catchers or unloading devices of the air systems (the first mode of operation of the fan) or the air flow into the dust-collecting channel can occur along it to avoid precipitation of impurities in it or to a certain working area where it is necessary to separate the forage grain from light impurities (second mode of operation of the fan). The third operating mode of the fan with the suction channel closed ensures that the air flow is only supplied to the delivery pipe.

1. Introduction

An important part of the technology of post-harvest grain processing is its purification from various weed impurities. The various grain cleaning machines equipped with open, closed or combined pneumatic systems, are used to perform these operations [1–8].

In most pneumatic systems of grain cleaning machines, centrifugal fans are used as an air flow generator. These fans do not provide a uniform airflow over the width of the pneumatic systems, which worsens the quality of cleaning the grain material. Therefore, as a generator of air flow, diametric fans are increasingly used to create a uniform airflow along the width of the pneumatic system, which is necessary for a more efficient grain cleaning [9–12].

Installation of fans of this type in the pneumatic system of grain cleaning machines in some cases requires the creation of guiding elements in front of its intake part. Therefore, with the development of universal grain cleaning machines, there is a need for more advanced designs of diametrical fans suitable for performing various functions in the air systems of these machines without the use of additional or special taps.

However, the use of a diametric fan with a suction channel located in the initial part of the spiral body and allowing it to be assembled into pneumatic systems of grain cleaning machines without additional or special taps is held back due to the lack of theoretical and experimental information in the scientific and technical literature.

2. Experimental

As an object of researches the experimental exemplar of the diametrical fan with the suction slot located in an initial part of a spiral body is chosen.

The diagram of a diametric fan with a suction channel located in the initial part of the spiral body is shown in Figure 1, and the general view of the experimental installation of this fan is shown in Figure 2.

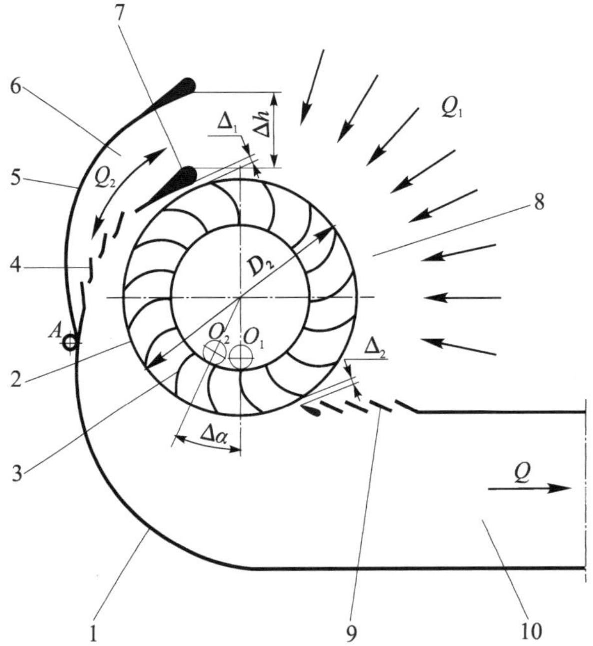

Figure 1. The diagram of a diametric fan with a suction channel located in the initial part of the spiral body, where 1 is spiral body; 2 is impeller; 3 are shovels; 4 is curved louvered grating; 5 is solid curved surface; 6 is suction channel; 7 is input edge; 8 is entrance window; 9 is straight louvre wall; 10 is discharge nozzle; → is airflow direction.

The diametric fan comprises a spiral body 1 with an inlet window 8 and a discharge nozzle 10 and a rectilinear louvre wall 9 separating them, as well as an impeller 2 with forward curved blades 3 installed in the body 1, the body 1 below the entrance edge 7 being provided with a curved louver grating 4, the louvers of which are directed from the impeller 2, and above the entrance edge 7 of the body 1 there is a pivotally fixed continuous curved surface 5 terminating on the fan body 1 at point *A* behind the curved louver the grating 4 and 6 forming the suction channel [13].

The fan impeller 2 with forwardly curved blades 3 was manufactured with an outer diameter D_2 equal to 0.3 m. The rotational speed of impeller 2 was 1060 min^{-1}. The width of the experimental setup was 0.1 m. At the output end of the discharge pipe of the experimental installation, replaceable dampers-diaphragms were installed, which made it possible to change the operating mode of the fan. To measure the air flow parameters in the flowing part of the experimental setup, a Pito-Prandtl tube, a micro manometer MMN and static pressure sensors were used.

In the investigation of the operation of a diametric fan with a suction channel located in the initial part of a spiral case, standard and particular techniques were used, using mathematical and physical modeling, modern instruments and computer technology with a package of programs for processing the results of experiments [14–16].

Figure 2. The general view of the experimental setup for studying the operation of a diametrical fan with a suction channel located in the initial part of the spiral body.

3. Results and Considerations

In operation of the diametric fan with the suction channel located in the initial part of the spiral body, the air flow is sucked through the inlet port 8 and injected into the rotary impeller 2, from which it again passes through the impeller 2 and is forced into the pipe 10. The body 1 forms and directs the air flow, coming off the impeller 2, and the wall 9 divides the incoming and outgoing air streams. In this case a large vortex is formed, the center of rotation O_1 of which is stabilized by a rectilinear louvre wall 9 and is located in the region of the radial clearance Δ_2 between the impeller 2 and the rectilinear louvre wall 9. With increasing load on the fan the center of rotation O_1 of the large vortex rotates by an angle $\Delta\alpha$. In addition, in the region of the radial clearance Δ_1, between the impeller 2 and the inlet edge 7 of the body 1, a small vortex is formed. As a result of the vacuum created by the rotary impeller 2, a small vortex is displaced along the direction of rotation of the impeller 2 towards the suction channel 6. As a result, additional air flow through the suction passage 6 to the fan casing 1 occurs. However, air sucking through the suction path 6 can occur until such a limit as the flow rate Q exceeds a certain flow limit $Q_{pred.}$ (the first operating mode of the fan), determined by the resistance of the network in which the fan is installed. If Q is less than $Q_{pred.}$ (the second mode of the fan operation), a positive static pressure will be observed in the suction channel 6, and air will be injected from the body 1 into the suction channel 6. This allows the fan to be used to perform various auxiliary operations as part of the grain cleaning machines. To transfer the diametric fan with the suction channel 6 to the normal (third) operating mode, the articulated solid curved surface 5 is rotated towards the curved louver grating 4, until it contacts the inlet edge 7. In the first and second modes of the fan operation, the total output Q of the diametric fan will be:

$$Q = Q_1 \pm Q_2, \tag{1}$$

where Q_1 is air flow through the inlet window of the diametrical fan, m^3/h; Q_2 is air flow through the suction channel of the diametrical fan, m^3/h.

The signs "plus" and "minus" in the equation (1) indicate whether the air flow is sucked or injected into the suction channel 6. The additional quantity of air Q_2 entering (emerging) through the suction channel depends on the area F of the curved louver grating, the coefficient μ live section, average speed υ_{sp} air and is:

$$Q_2 = \mu \cdot \upsilon_{sp} \cdot F. \tag{2}$$

Thus, knowing the average air flow rate υ_{sp} and the curved louver grating area F, it is possible to determine the amount of air Q_2 coming from the suction channel into the air main flow, or vice versa, the amount of air Q_2 exiting the suction path as the network resistance rises above a certain limit.

However, in practice it is rather difficult to apply equation (2), since it is difficult to measure the average velocity υ_{sp} of the air flow because of the limited space inside the suction channel and the turbulent movements of air near the curved louvre surface plates. It can be done more simply by using static pressure sensors installed in the flowing part of the suction channel.

Then the air flow Q_2 in the suction channel can be determined by the Equation [1, 2]:

$$Q_2 = \mu_c \cdot \Delta h \cdot B\sqrt{2 \cdot P_{sp,s\upsilon} / \rho_в}, \tag{3}$$

where μ_c is the flow coefficient, which depends on the Reynolds number Re and the shape of the hole, $\mu_c = 0.1\ldots0.5$; Δh is the depth of the suction channel located above the inlet edge of the casing of the diametrical fan, m; B is useful width of the diametric fan, m; $P_{sp,s\upsilon}$ is average static pressure in the suction channel of the fan, Pa; $\rho_в$ is density of air, kg/m^3.

In Equation (3) it is necessary to take into account that the largest coefficient μ_c refers to the smallest values of Δh, and the smallest μ_c refers to the largest values of Δh. In addition, the average static pressure $P_{sp,s\upsilon}$ should be taken in modulus, and the sign should be taken into account after calculation.

To determine the average velocity υ_{sp} of the air flow passing through the curvilinear lattice, we equate the Equations (2) and (3) and obtain a equation convenient for practical calculations:

$$\upsilon_{sp} = \frac{\mu_c \cdot \Delta h \cdot B\sqrt{2 \cdot P_{sp,s\upsilon} / \rho_в}}{F \cdot \mu}. \tag{4}$$

When determining the depth Δh of the suction channel in the zone of the inlet edge of the casing of the diametrical fan, the Equation (4) takes the form:

$$\Delta h = \frac{\mu \cdot \upsilon_{sp} \cdot F}{\mu_c \cdot B \cdot \sqrt{2 \cdot P_{sp,s\upsilon} / \rho_в}}. \tag{5}$$

Next, determine the average static pressure $P_{sp,s\upsilon}$ in the suction channel:

$$P_{sp,s\upsilon} = \left(\frac{\mu \cdot \upsilon_{sp} \cdot F}{\mu_c \cdot \Delta h \cdot B \cdot \sqrt{2 / \rho_в}} \right)^2. \tag{6}$$

It is not possible to analytically determine the value of Δh, since the values of the average speed υ_{sp} and the static pressure in the suction channel depend on the load on the diametric fan and are determined by its aerodynamic properties. Therefore, the rational value of Δh and its influence on the aerodynamic characteristics of the diametric fan has been studied experimentally.

Figure 3 shows the dependence of the air flow velocity υ_{sp} on the performance of a diametrical fan with a change in the depth of its suction channel Δh from 0.05 m to 0.15 m.

The analysis of the dependencies shows that they can be divided into a two-part conditional imaginary axis Q_{pred}: the "-Q_2" zone (the region of positive static pressures) and the zone "+Q_2" (the region of negative static pressures). In the «-Q_2» zone, the air flow is pumped from the fan casing to the suction channel, and from there to the atmosphere, while the fan flow is reduced by Q_2. This part of the dependencies graphically has the form of descending curves. In the "+Q_2" zone, phenomena opposite to the "-Q_2" zone occur and the dependencies have the form of linear functions. In general, analyzing Figure 3, it can be noted that the curves $\upsilon_{sp} = f(Q)$ are uniform and congruent to each other. This indicates the stability of the speed mode in the suction channel of the diametric fan.

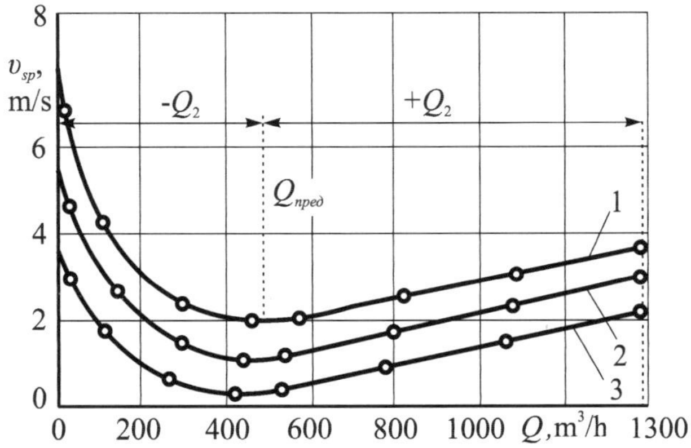

Figure 3. Dependences of airflow velocity υ_{sp} on the performance Q of a diametrical fan with a change in the depth of the suction channel, where 1 is at $\Delta h = 0.05$ m; 2 is at $\Delta h = 0.10$ m; 3 is at $\Delta h = 0.15$ m.

Thus, the highest airflow rates over the entire range of fan performance change are observed at $\Delta h = 0.05$ m. This is also confirmed by the dependence $\upsilon_{sp} = f(\Delta h)$ shown in Figure 4 with the maximum air flow rate $Q_{max} = 1300$ m^3/h and by the nominal flow rate $Q_н = 1070$ m^3/h.

Dependence analysis shows that when the depth of the suction channel Δh decreases, the air flow speed υ_{sp} increases and at the value $\Delta h = 0.05$ m reaches the maximum values. The interval of variation Δh from 0 to 0.05 m from the technical and economic point of view is inexpedient, since the design $\Delta h < 0.05$ m requires rather accurate manufacturing of the diametrical fan design, especially for grain cleaners with a wide air path.

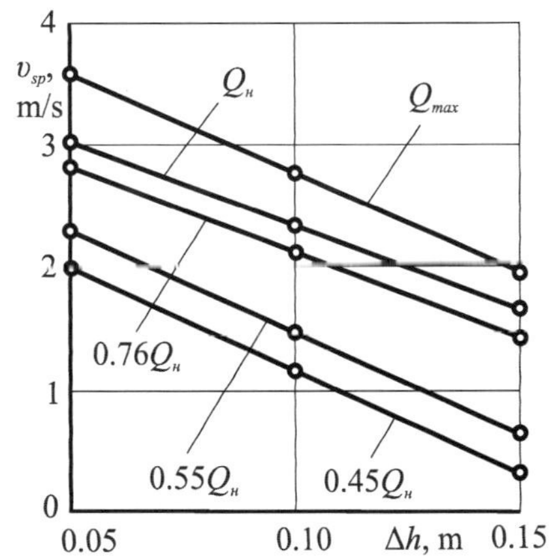

Figure 4. Dependences of the air flow speed υ_{sp} from the depth Δh of the suction channel of the diametrical fan at different capacities Q.

Investigation of the qualitative structure of the air flow in the delivery nozzle of a diametrical fan with a suction channel located in the initial part of the spiral housing showed that at values of $\Delta h = 0.05$; 0.10 and 0.15 m, the maximum rate variation coefficients γ, respectively, are 0.034, 0.067 and 0.078. This determines the stable operation of the fan.

4. Conclusion

Thus, on the basis of the conducted experimental studies of a diametric fan with a suction channel located in the initial part of the spiral body, it follows that the most rational value of the depth Δh of

IOP Conf. Series: Materials Science and Engineering **457** (2018) 012009 doi:10.1088/1757-899X/457/1/012009

the suction channel is 0.05 m or $\Delta h = 0.17D_2$. At the same time, changing the operating mode of the diametric fan, you can use the suction channel to perform various functions. When the diametric fan is operating in the $Q > Q_{pred}$ mode, it is possible to extract dusty air from the sedimentation chambers, dust collectors or unloading devices of the air systems without creating additional branches from the suction nozzle of the fan, thereby exporting a dusty air stream for repeated dust cleaning. When the diametric fan is operating in the $Q < Q_{pred}$ mode, it is possible to direct the air flow into the dust-collecting channel to avoid precipitation of impurities therein or to a certain work area where it is necessary to separate the feed grain from light impurities. With the suction port closed, the air flow is only supplied to the delivery pipe.

References

[1] Saitov V E 2012 *Innovations in post-harvest processing of grain material* (Saarbrucken: Lap Lambert Academic Publishing) (in Russian)

[2] Saitov V E *et al* 2012 *Investigation of processes in working bodies of grain separators* (Saarbrucken: Lap Lambert Academic Publishing) (in Russian)

[3] Blenk H *and* Trier H 1951 *Grundlagen der Landtechnik* **2** 17–25 Retrieved from URL: http://440ejournals.uni-hohenheim.de/index.php/Konstrukteur/article/view/1210/1053

[4] Wessel I 1963 *Grundlagen der Landtechnik* **18** 27-34 Retrieved from URL: http://440ejournals.uni-hohenheim.de/index.php/Konstrukteur/article/view/1212/1056

[5] Saitov V E *et al* 2013 *International Journal of Applied and Fundamental Research* **2** Retrieved from URL: http://www.science-sd.com/455-24419

[6] Murata S *and* Nisnihara K 1976 *Bulletin of JSME* **129(19)** 314-321

[7] *Mechanical Seed Cleaning and Handling Agriculture Handbook* **354** Retrieved from URL: https://naldc.nal.usda.gov/download/CAT87208718/PDF

[8] Schwanz H *and* Kutter W 1980 *Agratechnik* **30(11)** 495-497 Retrieved from URL: http://440ejournals.uni-hohenheim.de/index.php/agrartechnik/article/viewFile/5985/5691

[9] Coester R 1959 *Theoretische und experimentelle Untersuchungen an Querstromgeblasen* (Zürich: Mitteilungen aus dem für Aerodynamik) **28**

[10] Finkbeiner T 1966 *Landtechnischen Forschung* 16(3) 96-99 Retrieved from URL: http://440ejournals.uni-hohenheim.de/index.php/LTF/article/view/8833/8763

[11] Laakso H 1957 *Heizung-Luftung-Haustechnik* **8(12)** 324-325

[12] Furuno Y *et al* 2008 *Journal of the Japanese Society of Agricultural Machinery* **3(70)** 58-64

[13] Bolotov A K *et al* 2000 RF Patent 2156380 *Byull. Izobret.* **26** (in Russian)

[14] Zaydel A N 2009 *Errors in measuring physical quantities* (Moscow: Lan') (in Russian)

[15] Gmurman V E 2015 *Theory of Probability and Mathematical Statistics* (Moscow: Urait Publishing House) (in Russian)

[16] Leontiev V P 2014 *Work on the computer 2014: Windows 8.1* (Moscow: OLMA Media Group) (in Russian)

Determination technique of quasireference value of a carrier radio signal frequency of uncooperated onboard lighting source during radio monitoring

S F Boev[1], Y M Vikulova[2], I V Chebotar[3], M T Baldytchev[4], Y N Gaychuk[4], O V Galtseva[5]

[1]General Designer, JSC “RTI”, Moscow, 127083, Russia
[2]Project Manager, JSC “RTI”, Moscow, 127083, Russia
[3]Head of Research Department, Cherepovets Higher Military Engineering School of Radio Electronics, Cherepovets, 162622, Russia
[4]Postgraduate, Cherepovets Higher Military Engineering School of Radio Electronics, Cherepovets, 162622, Russia
[5]Associate Professor, School of Non-Destructive Testing & Security, National Research Tomsk Polytechnic University, Tomsk, 634034, Russia
E-mail: YVikulova@oaorti.ru

Abstract. The article discloses the technique defining quasireference value of a carrier radio signal frequency of uncooperated onboard lighting source by calculation and compensation of the Doppler shift at radio monitoring. Provided simulation results show checked operability of the offered technique with assessment of observation accuracy and received quasireference value of a carrier frequency accepted a radio signal compared with real sample under various conditions.

1. Introduction

Last years the theory and practice of complex application of radio monitoring and a radar-location in the conditions of the composite alarm and interfering situation a path of integration of land and air complexes of radio monitoring (RM) actively develops in spropercture of multiposition radar-tracking systems (radar station).

At the same time, in difference with classical multiposition radar station, the onboard radar complex (ORCOM) can be used as uncooperated integral lighting source for land and air complex of radio monitoring [1]. Uncooperated integral lighting source is understood in this article as an onboard radar complex, contact with which is not kept, and the operator of a land and air monitoring complex cannot exert impact on it. Parameters of such source signal and the characteristic of its carrier to a radio monitoring complex are not known in advance. The RM land and air complex is understood as a multiposition complex of radio monitoring composed of a ground station of management and processing (GSMP) and not less than four receiving positions – unmanned aerial vehicles (UAV-sensors) with designly placed onboard equipment of radio monitoring and controlled remotely with GSMP (Figure 1).

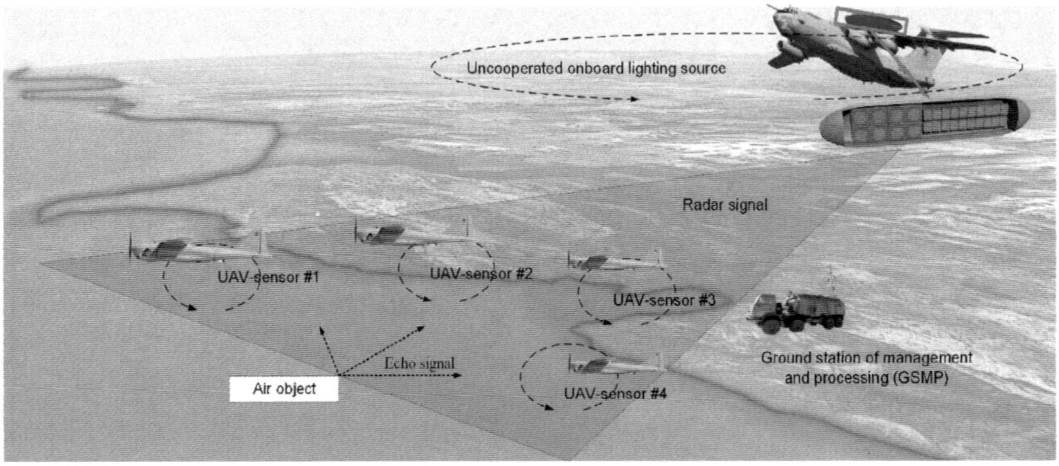

Figure 1. Structure of a multiposition complex of radio monitoring composed of a ground station of management and processing (GSMP) and not less than four receiving positions – unmanned aerial vehicles (UAV-sensors).

Implementation of similar fissile and passive system demands on the adaptation of devices of reception and processing of the RM complex to functioning in the conditions of dynamism where the received values of a carrier radio signal frequency can differ substantially from the proper in view of presence of the Doppler shift. The specified feature, in a consequence, complicates process of detection of the signals reflected from air objects with low power availability (echo signals) [2–6]. For the purpose of elimination of the designated feature within this article, there is presented the determination technique of quasireference value of a carrier frequency by the way of staying and compensation of the arising Doppler shift. The quasireference value of a carrier radio signal frequency of a source of integral lighting is understood within article as value of frequency of the considered radio signal taking into account compensation of the Doppler shift.

2. Determination technique of quasireference value of a carrier radio signal frequency of uncooperated onboard lighting source

The Doppler shift of a carrier frequency on a receiving position of the RM land and air complex (the UAV-sensor) is defined by expression:

$$f_{meas.} = f_{carr.}\left(1 \pm \frac{V_r}{c}\right), \tag{1}$$

where $f_{meas.}$ is the value of a carrier frequency of the probing signal measured on the UAV-sensor; $f_{carr.}$ is the proper value of a carrier frequency of the probing signal; V_r is the relative radial speed of the UAV-sensor and ORCOM carrier.

The technique includes nine sequentially carried-out stages.

2.1. Stage 1. Searching, detection and parameter estimation of a radio signal of integral lighting source (type of modulation, carrier frequency ($f_{carr.}$), signal duration (τ), time of arrival of a signal (t)). Each UAV-sensor having the panoramic receiver, on command with GSMP carries out searching of radio signals of ORCOM in the given frequency range.

2.2. Stage 2. Positioning of the ORCOM carrier by finding of the cross-correlation function of a delay of time of arrival of a radio signal for UAV-sensors (t_1, t_2, t_3, t_4), realizing at ground station of management and processing the difference and ranging system of a fixing.

For ensuring correct work of a difference and ranging way of a fixing of a source of a radio signal in space, the operator of GSMP defines the area of barraging and an echelon of the occupied heights for each UAV-sensor. Coordinates in real time are determined by signals of the block of navigation and time providing, the direction of driving and the real speed of each UAV-sensor, is carried out transfer of this information on GSMP where the high-precision binding to characteristic frame is carried out.

2.3. Stage 3. Calculation of radial speed of the UAV-sensor concerning the ORCOM carrier.

All angles in system are counted concerning one direction (chosen randomly on GSMP). When determining parameters of driving of the carrier of uncooperated source of integral lighting it is necessary to compensate the Doppler shifts, the bound to driving of receiving positions of the RM land and air complex.

Radial speed of the UAV-sensor concerning the ORCOM carrier can be found with that equation:

$$V_{r(UAV_i)} = V_{(UAV_i)} \cos\theta_i , \tag{2}$$

where $V_{(UAVi)}$ is value of the real speed of the UAV-sensor; θ_i is a projective angle of a vector of speed of the UAV-sensor to the line passing through the points characterizing location of number of the UAV-sensor and ORCOM carrier.

2.4. Stage 4. Compensation of the Doppler shift of a carrier frequency caused by driving of UAV-sensors.

Carrying out the mathematical calculation of values of the carrier signal frequency Doppler shift caused by UAV-sensors driving and calculate their compensation on an equation:

$$f_{comp.i} = f_i^{'}\left(1 \pm \frac{V_{r(UAVi)}}{c}\right) , \tag{3}$$

where $f_i^{'}$ is a carrier frequency of the probing signal on the corresponding UAV-sensor; $V_{r(UAVi)}$ is the radial speed of a receiving position concerning the ORCOM carrier.

2.5. Stage 5. Calculation of the intermediate values of parameters of driving of the ORCOM carrier.

Compensation of driving of UAV-sensors and expression projective angles of a vector of speed of the carrier integral lighting source on the corresponding straight lines 15, 25, 35, 45 through α_1, α_2, α_3, α_4, β_1, β_2, β_3, β_4, α, β, receive system from four equations with four unknowns α (roving), β (pitch), V (the real speed), $f_{carr.}$ (reference value of a carrier frequency of the probing integral lighting source signal).

$$\begin{cases} f_{comp.1} = f_{carr.}\left[1 - \dfrac{V\cos(\alpha_1 - \alpha)\cos(\beta_1 - \beta)}{c}\right] \\ f_{comp.2} = f_{carr.}\left[1 - \dfrac{V\cos(\alpha_2 - \alpha)\cos(\beta_2 - \beta)}{c}\right] \\ f_{comp.3} = f_{carr.}\left[1 - \dfrac{V\cos(\alpha_3 - \alpha)\cos(\beta_3 - \beta)}{c}\right] \\ f_{comp.4} = f_{carr.}\left[1 - \dfrac{V\cos(\alpha_4 - \alpha)\cos(\beta_4 - \beta)}{c}\right] \end{cases} . \tag{4}$$

Sequentially solve a set of equations (4) concerning V, α and β by a substitution method, receive equationes for finding of the corresponding sizes.

Express from the first equation of the f_{carr} system:

$$f_{carr.} = -\frac{f_{comp.1} \cdot c \cdot \left(1 - \frac{v \cdot \cos(\alpha - \alpha_2)\cos(\beta - \beta_1)}{c}\right)}{v \cdot \cos(\beta - \beta_1)\cos(\alpha - \alpha_2) - c}. \quad (5)$$

Expression (5) is substituted in the second equation of system (4) and express V_{12}:

$$V_{12} = \frac{c(f_{comp.2} - f_{comp.1})}{f_{comp.2} \cdot \cos(\beta - \beta_1) \cdot \cos(\alpha - \alpha_1) - f_{comp.1} \cos(\alpha - \alpha_2) \cdot \cos(\beta - \beta_2)}. \quad (6)$$

Expression (6) is substituted in the third equation of system (4) and expressed α:

$$\alpha' = \alpha_1 + \arctan\left(\frac{f_{comp1} \cdot \cos(\alpha_1 - \alpha_2) \cdot \cos(\beta - \beta_2) - f_{comp3} \cdot \cos(\alpha_1 - \alpha_2) \cdot \cos(\beta - \beta_2) - f_{comp1} \cdot \cos(\alpha_1 - \alpha_3) \cdot \cos(\beta - \beta_3)}{f_{comp1} \cdot \sin(\alpha_1 - \alpha_2) \cdot \cos(\beta - \beta_2) - f_{comp3} \cdot \sin(\alpha_1 - \alpha_2) \cdot \cos(\beta - \beta_2) - f_{comp1} \cdot \sin(\alpha_1 - \alpha_3) \cdot \cos(\beta - \beta_3) + f_{comp2} \cdot \sin(\alpha_1 - \alpha_3) \cdot \cos(\beta - \beta_3)} + \right.$$
$$\left. + \frac{+ f_{comp2} \cdot \cos(\alpha_1 - \alpha_3) \cdot \cos(\beta - \beta_3) - f_{comp2} \cdot \cos(\beta - \beta_1) + f_{comp3} \cdot \cos(\beta - \beta_1)}{f_{comp1} \cdot \sin(\alpha_1 - \alpha_2) \cdot \cos(\beta - \beta_2) - f_{comp3} \cdot \sin(\alpha_1 - \alpha_2) \cdot \cos(\beta - \beta_2) - f_{comp1} \cdot \sin(\alpha_1 - \alpha_3) \cdot \cos(\beta - \beta_3) + f_{comp2} \cdot \sin(\alpha_1 - \alpha_3) \cdot \cos(\beta - \beta_3)}\right) \quad (7)$$

Expression (7) is substituted in the fourth equation of system (4) and expressed β. As a result of transformations receive formulas for V, α and β on which find the intermediate values as there is an ambiguity of definition of angles α, β and the sign of speed of V.

2.6. Stage 6. Determination of the unique values of parameters of the ORCOM carrier movement.

The uniqueness of values of parameters of driving of the carrier of a source of integral lighting decides on next Equations:

$$V_{12} = \frac{c(f_{comp.2} - f_{comp.1})}{f_{comp.2} \cdot \cos(\beta - \beta_1) \cdot \cos(\alpha - \alpha_1) - f_{comp.1} \cos(\alpha - \alpha_2) \cdot \cos(\beta - \beta_2)}. \quad (8)$$

$$\alpha_{res.} = \alpha' + \left\{\frac{\pi \ at \ V > 0}{0 \ at \ V < 0}\right\}, \quad (9)$$

$$\beta_{res.} = \beta' + \left\{\frac{\pi \ at \ V > 0}{0 \ at \ V < 0}\right\}, \quad (10)$$

$$V_{res.} = \frac{|V_{12}| + |V_{13}| + |V_{14}| + |V_{34}| + |V_{23}| + |V_{24}|}{n}, \quad (11)$$

where $V_{res.} \in \{V_{12}, V_{13}, V_{14}, V_{23}, V_{34}, V_{24}\}$ n=6 at $f_{comp.1} \neq f_{comp.2}$, $f_{comp.1} \neq f_{comp.3}$, $f_{comp.1} \neq f_{comp.4}$, $f_{comp.2} \neq f_{comp.3}$, $f_{comp.2} \neq f_{comp.4}$, $f_{comp.3} \neq f_{comp.4}$.

2.7. Stage 7. Determination of radial speed of the ORCOM carrier concerning UAV-sensors.

On the basis of the received vector of speed of the carrier of a source of integral lighting determine its radial speed concerning each UAV-sensor:

$$V_{r(Ill.sour_i)} = V_{(Ill.sour_i)} \cos\alpha_i \cos\beta_i \ . \quad (12)$$

2.8. Stage 8. Definition of a carrier frequency of a radio signal of ORCOM.

On each UAV-sensor the carrier frequency of the accepted probing signal is defined by a formula:

$$f_{carr.i} = \frac{f_{comp.i}}{\left[1 \pm \frac{V_{R_{(Ill.sour_i)}}}{c}\right]}, \tag{13}$$

where $f_{comp.i}$ is the integral lighting source radio signal frequency measured on a receiving position taking into account compensation of the Doppler shift of the UAV-sensor caused by driving; $V_{R(Ill.sour.i)}$ is the radial speed of the carrier of a source of integral lighting relatively UAV-sensor.

2.9. Stage 9. Restitution of quasireference value of a carrier radio signal frequency of ORCOM.

For minimization of the error caused by distinction of measurement accuracy of parameters UAV-sensors of the RM land and air complex, the quasireference value of a carrier frequency of a radio signal of ORCOM is found how average value from calculated:

$$f_{carr.} = \frac{f_{carr.1} + f_{carr.2} + f_{carr.3} + f_{carr.4}}{4}. \tag{14}$$

The deviation of the received quasireference value of a carrier frequency of rather proper depends on accuracy of determination of parameters of driving of the carrier of a source of integral lighting and inspropermental accuracy of a frequency meter on the UAV-sensor.

3. Results and Discussion

For the purpose of operability check of the offered technique the imitation model is developed and consists of the software package with use of the LabVIEW-2014 environment. The model allows to imitate reception by ORCOM radio signal UAV-sensors (uncooperated integral lighting source) and to make error assessment of determination of required sizes a Monte-Carlo method.

Accuracy of finding the Doppler shift size is defined by potential accuracy of positioning and a vector of movement speed of the ORCOM carrier and also from accuracy of positioning of UAV-sensors and instrumental accuracy of a frequency meter on the UAV-sensor. At a simulation modeling as dynamically changeable sizes values of the relation signal/noise and parameters of a vector of speed of the ORCOM carrier and UAV-sensors were chosen. The value of a mean squared deviation (MSD) of positioning of ORCOM and UAV-sensors was accepted by a static fixed value.

Results of a simulation modeling allowed to construct schedules of mistake dependences in determination of carrier parameters of a source of integral lighting at various values of a ratio signal/noise on an entrance of the radio-receiving paths of the imitated RM complex (Figure 2 and 3). Researches were conducted on condition of impact on the radio signal RM complex with the linear - the frequency modulation with the f_{carr} parameters. = 1 GHz, $f_{dev.}$ = 250 MHz, τ = 1 ms and the real speed of the carrier of a source of integral lighting of V = 205 m/s.

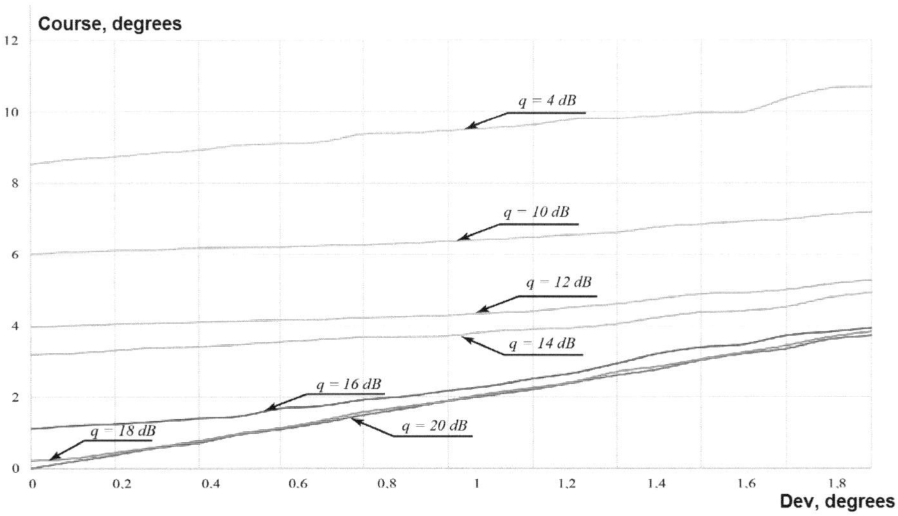

Figure 2. Dependence of an error of definition of the direction of driving of the carrier of a source of integral lighting on positioning MSD at various ratios signal/noise on an entrance of the radio-receiving path of the imitated RM complex.

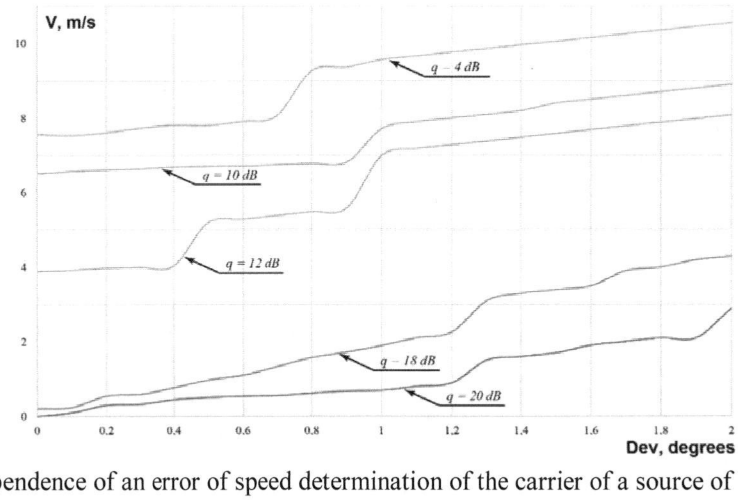

Figure 3. Dependence of an error of speed determination of the carrier of a source of integral lighting on positioning MSD at various ratios signal/noise on an entrance of the radio-receiving path of the imitated RM complex.

Schedules show operability of the offered technique in the conditions of a ratio signal/noise from 20 to 4 dB. Assessment of accuracy of obtaining quasireference values of a carrier frequency of the probing signal from the proper showed that depending on the relation signal/noise elimination of fluctuations, the bound to a Doppler effect, is possible ranging from 86 up to 97%.

4. Conclusion

The article discloses the technique defining quasireference value of a carrier radio signal frequency of uncooperated onboard lighting source by calculation and compensation of the Doppler shift at radio monitoring. Provided simulation results show checked operability of the offered technique with assessment of observation accuracy and received quasireference value of a carrier frequency accepted a radio signal compared with real sample under various conditions. It is established that depending on the relation signal/noise elimination of fluctuations, the bound to a Doppler effect, is possible ranging from 86 up to 97%.

References

[1] Boev S F *et al* 2017 *Journal of Radio Electronics* **4** 1–17 URL: http://jre.cplire.ru/jre/apr17/8/text.pdf (in Russian)

[2] Tuteur F B 1965 *Proc of the IEEE* **53(5)** 487-488

[3] Sheehy J J 1978 *Journal of the Acoustical Society of America* **63(1)** 81-90

[4] Kassam S A 1980 *IEEE Transactions on Information Theory* **II–26(5)** 595–602

[5] Scharf L L and Lytle D W 1971 *IEEE Transactions on Information Theory* **II–17(4)** 404–439

[6] Singer R A *et al* 1974 *IEEE Transactions on Information Theory* **IT-20** 423-432

IOP Conf. Series: Materials Science and Engineering **457** (2018) 012011 doi:10.1088/1757-899X/457/1/012011

Investigation of the speed regime of tractor diesel engine running on natural gas with recirculation

V A Likhanov[1] and O P Lopatin[2]

[1]Professor, Vyatka State Agricultural Academy, Kirov, Russia
[2]Associate Professor, Vyatka State Agricultural Academy, Kirov, Russia
E-mail: nirs_vsaa@mail.ru

Abstract. The work is devoted to the use of compressed natural gas as an alternative motor fuel to reduce the toxicity of exhaust gases of tractor diesel engines. At the same time, the urgency of using natural gas in combination with exhaust gas recirculation (EGR) (EGR was used to reliably reduce nitrogen oxides increased as a result of the use of natural gas) in diesel was substantiated. The paper presents the results of experimental studies on the effect of natural gas and EGR on the effective characteristics, combustion process indicators, the toxicity and smokiness of exhaust gases depending on the change in the speed of the crankshaft of the tractor diesel. On the basis of the results of the study, it is proposed to maintain the following ratio: natural gas – 80%, the fuse portion of diesel fuel – 20%. When working on natural gas with EGR, the regularity of supply of recirculated gases from 20 to 40% depending on the mode of operation of the tractor diesel is established. The paper defines and presents the optimal values of the effective parameters, characteristics of the combustion process, toxicity and smoke exhaust gases when working tractor diesel 4F 11.0/12.5 on natural gas with EGR. The translation of the specified diesel fuel oil to natural gas helped to reduce the contents in the exhaust gases when working with EGR 20% of the exhaust gas of nitrogen oxides by 43.2%, of carbon black in a 5.6 times that of carbon dioxide 33.3%, carbon monoxide by 10.0%.

1. Introduction

The steady increase in the consumption of mobile oil fuel and the constant tightening of environmental requirements for it compels mankind to seek alternative energy sources, since their use should reduce the environmental damage caused to the environment, as a result of the use of oil fuels [1–4].

Reduction of the share of transport in pollution is one of the main state priorities defined by the Transport strategy of the Russian Federation till 2030 approved by the order of the Government of the Russian Federation No 1734-r of 22.11.2008. In order to reduce the negative impact of transport on the environment, it is planned to develop and implement mechanisms of state regulation that motivate the transfer of vehicles to environmentally friendly fuels. Currently, among all modes of transport, the problem of increasing the environmental purity of exhaust gases is most acute for mobile agricultural machinery. The increasing relevance of this problem for tractors and other agricultural machinery equipped with diesel engines is due to the existing and planned introduction of emission control standards for toxic exhaust components [5–9].

In the exhaust gases of a tractor diesel engines running on fuel oil that contains more than 200 toxic components. One of the most harmful are nitrogen oxides and solid particles consisting mainly of soot. Of all the variety of existing methods to reduce the toxicity of exhaust gases, the greatest effect can be achieved by the use of alternative fuels that are less prone to soot formation in the combustion

IOP Conf. Series: Materials Science and Engineering **457** (2018) 012011 doi:10.1088/1757-899X/457/1/012011

chamber due to the characteristics of the chemical composition and at least physical properties. First of all, natural gas can be attributed to such fuels [10–14].

2. Experimental

In Vyatka state agricultural Academy on the basis of the Department of thermal engines, Vehicles and tractors experimental researches of tractor diesel 4F 11.0/12.5 for work on natural gas with EGR were carried out [15].

In order to determine and optimize the basic parameters of the diesel when working on compressed natural gas with EGR, its bench studies on diesel, gas-diesel and gas-diesel processes with EGR were carried out. It was found that diesel is stable on compressed natural gas at a ratio to nominal conditions: gas – 80%, the ignition portion of diesel fuel – 20%. Further study of the process was carried out in this ratio.

The experimental setup consisted of the SAK-N670 electrothermal stand with the balancing pendulum mechanism (Figure 1), the tractor diesel 4F 11.0/ 12.5 equipped with the exhaust gas recirculation system (Figure 2), the measuring equipment.

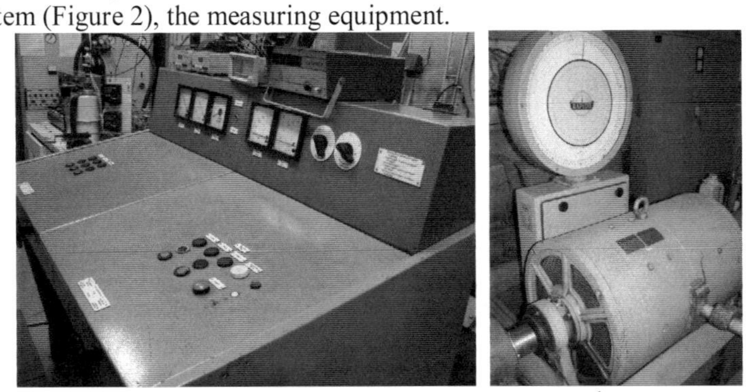

a b

Figure 1. Electroforming stand, where a is stand; b is balance pendulum machine.

a b

Figure 2. General view of the engine mounted on the stand, where a is view of the EGR system; b is view of the mixer-dispenser of natural gas and recycled gas supply.

3. Results and Considerations

According to power and economic performance of a tractor diesel engine 4F 11.0/12.5, depending on the change of the rotation frequency without regulatory branches is presented in Figure 3.

Analyzing the curves corresponding to the operation of the diesel on natural gas, it can be noted that the increase in the rotational speed leads to an increase in the hourly fuel consumption of G_F, the fuse portion of the diesel fuel G_{DF}, the effective power of N_e, the hourly air consumption of G_a and the temperature of the exhaust gases t_r. The value of the specific effective fuel consumption is g_e minimal at a speed of 1700 min^{-1} and is 198 g/(kW·h). The value of the effective coefficient of efficiency η_e maximum at a rotational speed of 1700 min^{-1} and is 0.37. Torque M_t maximum at speed of 1700 min^{-1}

and is 259 N·m. The Coefficient of excess air α with increasing frequency of rotation decreases slightly. Analyzing the operation schedules of the natural gas diesel with EGR, it is clear that when the gas diesel with EGR 10% operates, the effective power and torque remain practically at the level of the gas-diesel process without EGR. By increasing the degree of EGR up to 20% there is a decrease in effective power and torque. Thus, the use of EGR 20% at n=1400 min^{-1} results in a 2.5% drop in the effective power of the gas-diesel process without EGR. When working on gas-diesel process by increasing the frequency of rotation in the range from 1400 to 2400 min^{-1} g_e varies from 204 to 213 g/(kW·h), when operating on natural gas with 10 and 20% EGR de varies from 207 to 218 g/(kW·h) and from 214 to 224 g/(kW·h), respectively.

Thus, the application of 10 and 20% EGR in the range n from 1400 to 2400 min^{-1} results in an increase of g_e to 1.5÷2.4% and up to 4.9÷5.2%, respectively. The effective efficiency factor, the maximum of which is achieved at a frequency that corresponds to the maximum torque, for the gas-diesel process is 0.37, and for gas – diesel with EGR 10 and 20% - 0.36 and 0.35, which is lower by 2.7 and 5.4% respectively. During the operation of the gas diesel with exhaust gas recirculation, the hourly air flow rate and the ratio of the increase in the EGR degree decrease. The exhaust gas temperature when operating with the recirculation decreases slightly with increasing degree of recycling. The temperature of the EGR gases at work with EGR 20% at all speed modes does not exceed 67°C.

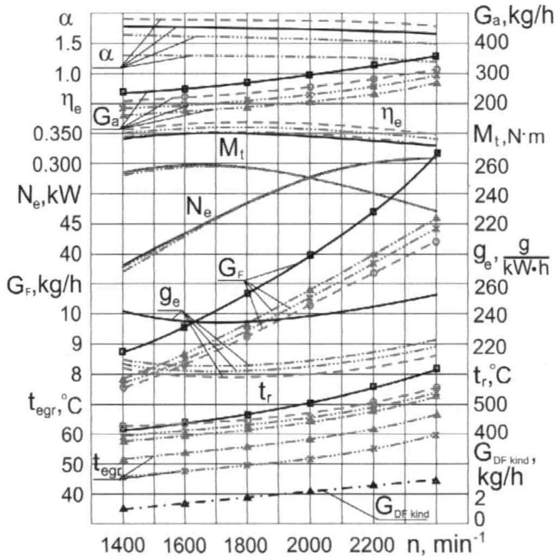

Figure 3. The impact of the use of natural gas and exhaust gas recirculation on power and economic performance of a diesel 4F 11.0/12.5 based on of changes in the frequency of the crankshaft at Θ_{inj} = 23°, where ——— is diesel fuel; — — — is gas-diesel fuel; ···—···— is gas-diesel fuel with EGR 10%; —··—··— is gas-diesel fuel with EGR 20%; ▲· — ·▲ is consumption of kindle diesel fuel.

The content of toxic components in the exhaust gases of the tractor diesel 4F 11.0/12.5 depending on the change in the rotational speed is shown in Figure 4.

Analyzing charts of the content of toxic components when working on gas-diesel and gas-diesel processes with EGR, it can be noted that the nature of the curves corresponds to the diesel process, while only their numerical values change. Thus, when working on a gas-diesel process with a EGR 20% with an increase in the rotational speed from 1400 to 2400 min^{-1}, the content of nitrogen oxides (NO_x) decreases from 900 to 700 ppm, or by 22.2%, which is lower than the diesel process by 30.8 and 30.0% respectively. The content of total hydrocarbons increases both with increasing rotational speed and with the use of natural gas. Thus, the use of natural gas in the range of rotational frequencies from 1400 to 2400 min-1 leads to an increase in the total hydrocarbon content (CH_x) in the exhaust gases by 6.8 and 4.0 times respectively.

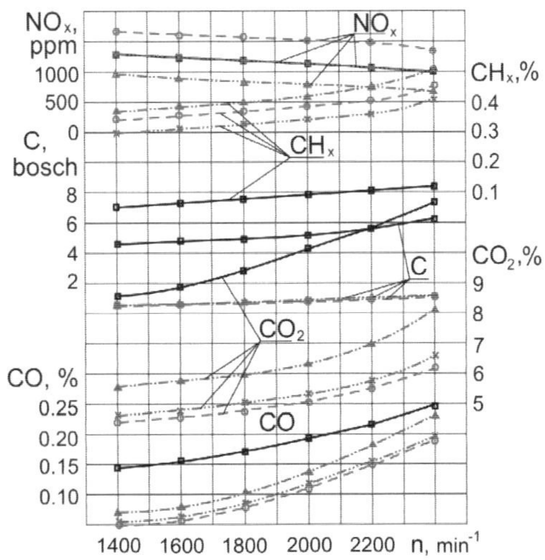

Figure 4. The impact of the use of natural gas and the EGR on the content of toxic components in the exhaust gases of a tractor diesel engine 4F 11.0/12.5 based on of changes in the frequency of the crankshaft at $\Theta_{inj} = 23^o$, where ——— is diesel fuel; — — — is gas-diesel fuel; ···—···— is gas-diesel fuel with EGR 10%; —··—··— - gas-diesel fuel with EGR 20%.

The use of EGR 10% in the gas diesel results in a decrease of 11.8% and 9.1%, respectively. But the use of EGR 20% on the gas diesel causes an increase in the CH_x content in the exhaust gases by 8.8 and 15.9% for the rotation frequencies of 1400 to 2400 min^{-1}, respectively. Increasing the rotational speed causes an increase in the carbon monoxide (CO) and carbon dioxide (CO_2) content of the exhaust gases. The use of natural gas causes a decrease in the content of oxide and carbon dioxide, and with an increase in the EGR, the content of CO and CO_2 increases again.

But still, when the gas diesel with EGR 10% in the range of rotation frequencies from 1400 to 2400 min^{-1} decreases by 2.4 times and by 20.8% respectively, and a decrease in CO_2 by 46.0 and 43.6% in relation to the diesel process. Analyzing the content of soot (C) in the exhaust gases, we see that with an increase in the rotational speed, the soot content in the exhaust gases is increased both in the gas-diesel and in the gas-diesel processes with the EGR. The use of natural gas significantly reduces the soot content of the exhaust gases, and the use of EGR leads to an increase in the soot content, and with an increase in the EGR, the soot content increases. Thus, when the gas diesel operates in the range of rotation frequencies from 1400 to 2400 min^{-1}, the particulate content decreases by 8.8 and 5.1 times, respectively, in relation to the diesel process. The use of the same EGR 10% causes an increase in particulate matter, which does not exceed 10%.

Figure 5 shows the combustion process parameters of the tractor diesel 4F 11.0/12.5 depending on the change in speed.

Considering the operation of the diesel engine on natural gas, it is clearly seen that with an increase in the rotational speed there is a decrease in the maximum averaged temperature of the gases in the cylinder (T_{max}), maximum pressure (p_z), the degree of pressure increase (λ), the rigidity of the combustion process $(dp/d\varphi)_{max}$ and an increase in the angle corresponding to the ignition delay period (φ_i). Thus, in the speed range from 1200 to 2400 min^{-1}, T_{max} decreases from 3100 to 3000 K, or by 3.2%; p_z max from 10.9 to 8.1 MPa, or by 25.7%; λ from 2.56 to 1.94, or by 24.2%; $(dp/d\varphi)_{max}$ 0.98 to 0.61 MPa/deg, or 37.8%, and an increase of φ_i from 19.0° to 33.5° of crankshaft rotation, or 14.5° of crankshaft rotation. In the case of natural gas diesel with EGR, the nature of the curves coincides with the gas-diesel process completely, and the values of the graphs "lie" below the curves of the gas-diesel process (except φ_i) throughout the speed range and decrease with an increase in the EGR. Thus, when working on a gas-diesel process with a EGR 10% in the range from 1200 to 2400 min^{-1}, there is a decrease in the maximum averaged temperature by 200 K in comparison with the gas-diesel process, the maximum pressure by 7.3–4.9%, the degree of increase in the pressure λ by 6.3–5.3%, the rigidity of the combustion process by 18.4–8.2% and an increase in the angle φi from 0 to 0.5°. The application of the same EGR 20% in the range from 1200 to 2400 min^{-1} leads to a decrease in relation to the gas-diesel process of the maximum averaged temperature by 350–400 K, the maximum pressure by 11.0–13.6%, the degree of increase in the pressure λ by 10.2–12.4%, the rigidity of the combustion process by 33.7–11.5% and an increase in the angle from 0.5 to 2.5°.

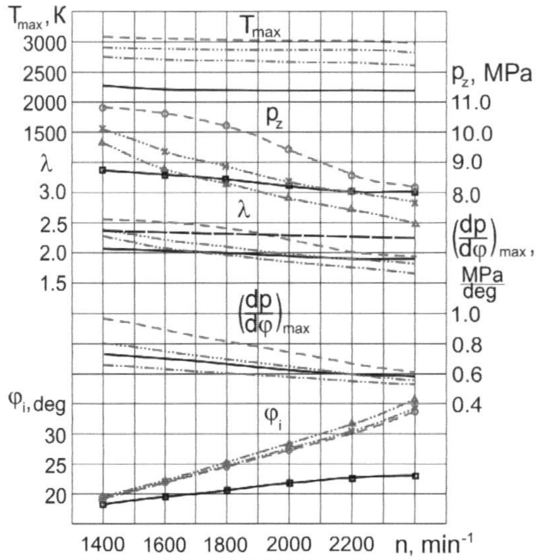

Figure 5. The impact of the use of natural gas and EGR on the performance of the combustion process of a diesel engine 4F 11.0/12.5 based on of changes in the frequency of the crankshaft at $\Theta_{inj} = 23°$, where ——— is diesel fuel; — — — is gas-diesel fuel; ···—···— is gas-diesel fuel with EGR 10%; —··—··— is gas-diesel fuel with EGR 20%.

The use of EGR at the gas diesel 4F 11.0/12.5 makes it possible to effectively reduce the toxicity of exhaust gases by nitrogen oxides. The main issue to be solved when equipping the gas diesel with a EGR system is the development of an automatically controlled drive that changes the law of control of the exhaust gas bypass depending on the mode of operation of the gas diesel. The dependence of the change in the EGR from the load of the tractor diesel 4F 11.0/12.5 when operating on natural gas is shown in Figure 6.

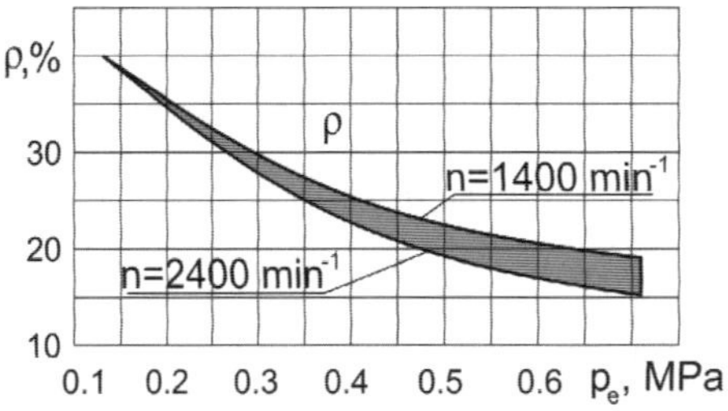

Figure 6. The dependence of the degree of EGR gas diesel 4F 11.0/12.5 from load.

For all frequencies of rotation of a cranked shaft of the tractor diesel engine increasing the EGR for different load modes leads to a larger decrease in the concentration of nitrogen oxides in the exhaust gases.

4. Conclusion

1. On the basis of the results of researches of tractor diesel 4F 11.0/12.5 it is offered to support the following ratio: natural gas - 80%, the fuse portion of diesel fuel - 20%.

2. When working on natural gas with EGR, the regularity of supply of recirculated gases from 20 to 40% depending on the mode of operation of the tractor diesel is established.

3. The possibility of maintaining the power parameters of the tractor diesel 4H 11.0 / 12.5 when operating on natural gas with EGR at the level of serial diesel: N_e=55.2 kW (n=2200 min^{-1}), N_e=46.5 kW (n=1700 min^{-1}).

4. The values of combustion process parameters are determined. The nominal speed mode: diesel process - T_{max}=2190 K; p_z=8.1 MPa; λ=1.90; $(dp/d\varphi)_{max}$=0,59 MPa/deg; φ_i=22.5°; gas-process - T_{max}=3010 K (increase by 37.4%); p_z=8.5 MPa (increase by 4.9%); λ=2.0 (increase by 5.3%); $(dp/d\varphi)_{max}$=0.69 MPa/deg (increase by 17.0%); φ_i=30,0° (increase by 33.3%); gas-process with EGR 10% - T_{max}=2790 K (increase by 27.4%); p_z=8.1 MPa (corresponds to diesel fuel); λ=1,9 (corresponds to diesel fuel); $(dp/d\varphi)_{max}$=0.60 MPa/deg (increase by 1.7%); φ_i=31.0° (increase by 37.8%); gas-process with EGR 20% - T_{max}=2680 K (increase by 22.4%); p_z=7.5 MPa (decrease by 7.4%); λ=1.8 (decrease by 5.3%); $(dp/d\varphi)_{max}$=0.54 MPa/deg (decrease by 8.5%); φ_i=32.0° (increase by 6.7%). Max torque: diesel process - T_{max}=2210 K; p_z=8.6 MPa; λ=2.0; $(dp/d\varphi)_{max}$=0.64 MPa/deg; φ_i=20.0°; gas-process - T_{max}=3050 K (decrease by 38.0%); p_z=11,0 MPa (decrease by 27.9%); λ=2.6 (decrease by 30.0%); $(dp/d\varphi)_{max}$=0,83 MPa/deg (decrease by 29.7%); φ_i=22.0° (decrease by 10.0%); gas-process with EGR 10% - T_{max}=2880 K (decrease by 30.3%); p_z=10.2 MPa (decrease by 18.6%); λ=2.40 (decrease by 20.0%); $(dp/d\varphi)_{max}$=0.76 MPa/deg (decrease by 18.8%); φ_i=24.0° (decrease by 20.0%); gas-process with EGR 20% - T_{max}=2710 K (decrease by 22.6%); p_z=9.2 MPa (decrease by 7.0%); λ=2,32 (decrease by 16.0%); $(dp/d\varphi)_{max}$=0.66 MPa/deg (decrease by 3.1%); φ_i=23.0° (decrease by 15.0%).

5. The translation of the specified diesel fuel oil to natural gas helped to reduce the contents in the exhaust gases when working with EGR 20% of nitrogen oxides by 43.2%, of carbon black in a 5.6 times that of carbon dioxide 33.3%, carbon monoxide by 10.0%.

References

[1] Ter-Mkrtichyan G G *et al* 2017 *Pollution Research Paper* **36(3)** 678–683.
[2] Kholod N *and* Evans M 2016 *Environmental Science & Policy* **56** 1-8 doi: 10.1016/j.envsci.2015.10.017
[3] Arat H T *et al* 2016 INT J HYDROGEN ENERG **41(19)** 8354–8363 doi: 10.1016/j.ijhydene.2015.11.183
[4] Lazarev E *and* Lomakin G 2014 *WIT Transactions on Ecology and the Environment* **190(1)** 677–683
[5] Ugay S M *et al* 2014 *Life Science Journal* **11(10s)** 515-517
[6] Gough R V and Bruno T J 2012 *Energy and Fuels* **26(11)** 6905–6913
[7] Yang L *et al* 2014 *Renewable and Sustainable Energy Reviews* **40** 1133–1152
[8] Khan M I *et al* 2016 *Renewable and Sustainable Energy Reviews* **66** 702–741
[9] Zgodziński T 2012 *Bezpieczenstwo i Technika Pozarnicza* **26** 53–66
[10] Mikulski M *and* Wierzbicki S 2016 *Journal of Natural Gas Science and Engineering* **31** 525–537
[11] Chai X *et al* 2016 *Progress in Energy and Combustion Science* 56 33–70
[12] Sanjid A *et al* 2013 *Renewable and Sustainable Energy Reviews* **27** 664–682
[13] Meyer D D *et al* 2014 *International Biodeterioration & Biodegradation* **95** 356–363
[14] Azizi Z *et al* 2014 *Chemical Engineering and Processing* **82** 150–172
[15] Likhanov V A *and* Lopatin O P 2017 *Engineering journal: Science and innovation* **1(61)** doi: 10.18698/2308-6033-2017-1-1573

IOP Conf. Series: Materials Science and Engineering **457** (2018) 012012 doi:10.1088/1757-899X/457/1/012012

Determination of optimal technological parameters of a washing machine on the basis of its energy studies

A V Mayorov[1] and N E Yaytseva[2]

[1]Associate Professor, Mari State University, Yoshkar-Ola, Russia
[2]Student, Mari State University, Yoshkar-Ola, Russia
E-mail: natalya.yaytseva@mail.ru

Abstract. It is necessary to study supplied power of all modes for more precise determination of optimal mode of tin cans surface washing. The following parameters are defined: effective power that is expended only for washing, effectiveness factor of power use. Dependence of power expenditure, energy output and effectiveness factors of power use on rotational rate of the washing machine' driven wheel are determined. Following the results of research, optimal technological parameters of the washing machine are determined. According to the results of research, the following optimal parameters are defined: necessary cleaning quality with optimal rotation rate of the driven wheel can be reached with temperature (74...85°C) and concentration of washing solution of 1.7...3 g/l. If it is necessary to increase production capacity from 88 to 100 cans per minute, it is important to increase rotation rate of the driven wheel from 31 to 35 r/min. Concentration of solution should be minimal due to optimal rotation rate and constitute c = 1.7 g/l. Energy output also increases by 2% that is not significant increase in comparison with increase of machine production capacity. Quality of cans surface cleaning will be K ≥ 93 % with temperature from 72 to 85 °C.

1. Introduction

Quality of cans cleaning, labour capacity, industrial standards, and cost price of canned food production depend on excellence of technology and washing machines.

A large number of modern washing machines work in an irrational regime and are energy and metal consuming, so their use is unprofitable. Consequently, the development of new efficient machines and the improvement of existing washing machines in the canning industry is great reserve to reduce energy consumption, materials and cost of the whole production process of canned food.

Therefore, research aimed at improving the washing process filled cans in the production of canned food are very relevant and have great importance in the food industry.

Appliance of submersible cleaning as the less energy-demanding method is more prospective according to the criteria of specific energy costs on a unit of the surface being cleaned.

When the washing machine is used, the same degree of cleaning of the products can be obtained with different modes (for different combinations of factor values). As the speed of the wheels increases, the intensity of the process increases. However, the costs also increase - the required power, the pressure in the bearings increase, that is, the operating mode of the machine deteriorates from the point of view of force analysis. Therefore, the main goal of experimental research is the determination of the optimum technological parameters of the washing machine.

IOP Conf. Series: Materials Science and Engineering **457** (2018) 012012 doi:10.1088/1757-899X/457/1/012012

The optimization task was the choice of such a combination of the values of the investigated factors, which ensure the quality of treatment required by the standard, at which energy consumption will be the least.

2. Experimental

It is necessary to study supplied power of all modes for more precise determination of optimal mode of tin cans surface washing [1–3].

Let us consider power structure and determine effectiveness. To get cleaning physical impact of washing fluid on the object in submerged washing machines, it is necessary to coerce rotational movement to the wheels with the help of which eccentric motion of the objects being washed is carried out. Energy necessary for this process is consumed on rotation of the wheels, displacement of the objects being washed, on passage of washing fluid' hydrodynamic resistance to the wheels rotation and the objects being washed, and on passage of friction force in bearings. Because of the comparative low frictional force in the bearings of the examined machine, they were not taken into account.

Only energy consumption for passage of hydrodynamic resistance of washing fluid to movement of the objects being washed is useful among above listed expenditures [4–7]. Effective power mostly expended on cleaning (fluid physical impact on contamination) is represented by the following Equation:

$$N_{ef} = N_{ttl} - N^{fl}_{whl} - N_{whl.can} + N_{whl} \tag{1}$$

where N_{ttl} is total output consumed on wheel rotation and object being washed in washing fluid; N^{fl}_{whl} is power expended on wheel rotation without cans displacement in fluid; $N_{whl.can}$ is power expended on wheel rotation and cans displacement in the empty bath; N_{whl} is power expended on wheel rotation without cans displacement in bath without washing fluid.

Then effectiveness factor of power use can be found from the Equation

$$\eta = N_{ef} / N_{ttl} \tag{2}$$

The meter Mercury 230 AR-01R3 was used to measure required power of washing machine drive and its components.

On the Figure 1 there are graphs of power and its components change depending on rotation rate of the driven wheel of the washing machine. These graphs are based on the results of power in three drive electromotor phases by the meter Mercury 230 AR-01R3 in studied modes during washing of tin can surface and their continuous displacement.

3. Results and Considerations

An empirical dependence between expended power and rotation rate of the driven wheel with approximation validity of R^2=0.9963 was defined on a personal computer with the help of the programme Excel after results processing of power measurement in three drive electromotor phases:

$$y = 204.36 \cdot e^{0.1426 \cdot x} \tag{3}$$

On the basis of the graph (Figure 1) it may be concluded that with increase of the driven wheel' rotation rate quality of cleaning and energy consumption also increase. Energy consumption increases more intensively as dependence is exponential. That is why there should be an optimal mode of the washing machine (in relation to energy consumption) to achieve technologically planned cleaning quality.

The optimal mode of the washing machine was defined on the basis of received relevant regression equation of cleaning quality depending on studied parameters from the plan of three-factor experiment. It shows that power from 361.5 to 417 W corresponding to 30 and 35 r/min is necessary to receive cleaning quality relevant to the standard. Graphs on the Figure 1 show that damaging components of expended power (N_{whl}, N^{fl}_{whl}, $N_{whl.can}$) increase with increase of rotation rate n, but less intensively than total output (N_{ttl}).

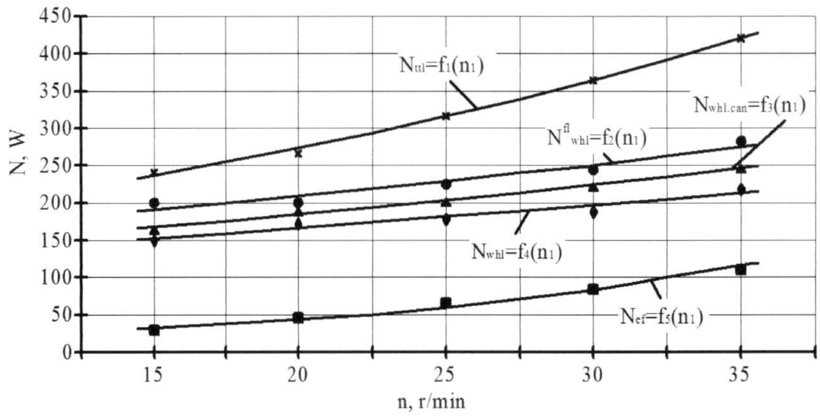

Figure 1. Dependence of energy consumption on rotation rate of the washing machine' driven wheel Power, where N_{ttl} – total, N^{fl}_{whl} – power expended on wheel rotation without cans displacement in fluid, $N_{whl.can}$ – power expended on wheel rotation and cans displacement in the empty bath, N_{whl} – power expended on wheel rotation without cans displacement in bath without washing fluid, N_{ef} – effective.

That is why effective (useful) power (N_{ef}) found with the help of the equation (1) and expanded mostly on passage of hydrodynamic pressure increases quite intensively. Relation of this effective power to total output, according to the equation (2), shows effectiveness of power use.

Effectiveness factor of power use of the washing machine can be considered according to the graph (Figure 2).

The analysis of the results shows that effectiveness factor in the optimal mode of the washing machine with eccentric motion of the objects being washed and fluid turbulization by air barbotage from the outer side of the cleaned objects is η=0.23…0.26. For comparison: the factor of submersible washing machines of vibration type is η = 0.1…0.2, and it is η = 0.05…0.10 in jet-type washing machines [1].

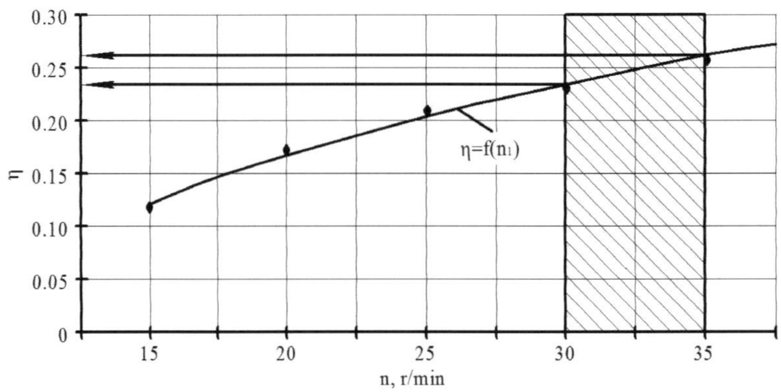

Figure 2. Dependence of effectiveness factor η of power use on rotation rate n of the driven wheel.

Research showed that equal quality of cans surface cleaning can be achieved due to different combinations of values of studied factors. That is why the optimization task is to find such a combination of these values due to which energy consumption would be at a minimum level.

Required power is found from the dependence of expanded power on rotation rate of the driven wheel $N_{ttl} = f_1(n_1)$ (Figure 1).

On the Figure 3 there are received values of machine capacity depending on rotation rate of the driven wheel found theoretically and experimentally. W_{exp} is defined on the basis of data represented in the Table 1.

Table 1. Results of capacity measurement due to different rotation rate of the driven wheel

Rotation Rate of the Wheel n_1, r/min^{-1}	Machine Capacity W, pcs/min					Arithmetic Mean Value W, pcs/min
	Repetitions					
	1	2	3	4	5	
15	45.55	45.11	45.39	44.54	45.87	45.21
20	54.55	57.53	56.08	60.48	62.02	57.99
25	75.42	75.14	75.99	76.58	78.74	76.36
30	84.69	88.82	91.74	89.22	82.25	87.21
35	98.68	99.59	93.53	97.72	100.84	98.01

Therefore, experiments approved validity of theoretical measurement.

From the graph (Figure 3) it can be seen that machine capacity increases on empirical dependence with increase of rotation rate of the driven wheel

$$y = 13.48 \cdot x + 32.519. \tag{4}$$

Energy output can be found with the help of the Equation (5).

$$E = N/60 \cdot W, \tag{5}$$

where N is required power, W; W is machine capacity, n/min; E is energy output of washing process, W·h/pcs.

The least expenditures for achieving planned quality level of cleaning and factors parameters providing these expenditures are found after solving this equation on minimum for different combinations of values.

On the Figure 4 there is a graph of energy output change depending on rotation rate of the driven wheel. Minimal energy consumption corresponds to the optimal value of rotation rate n_{opt1}. In this case $E_{min} = 68.9765 \cdot 10^{-3}$ W·h/pcs, $n_{opt1} = 30.6$ r/min.

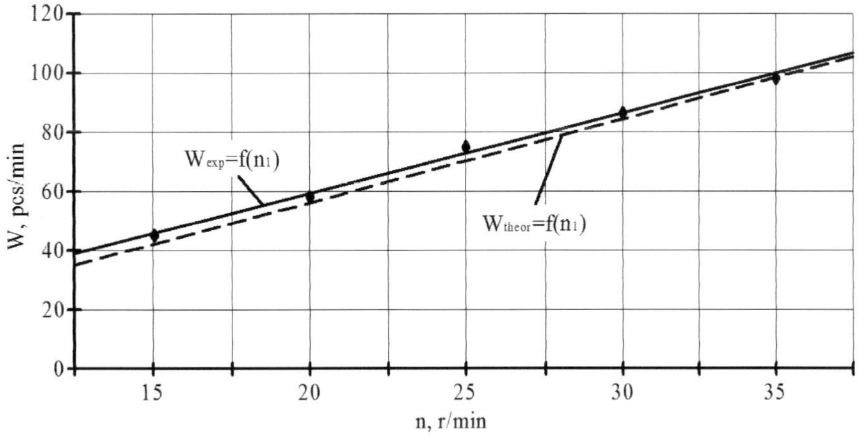

Figure 3. Theoretical (– – – –) and experimental (———) dependence of machine capacity on rotation rate of its driven wheel.

Productive capacity of the washing machine for the cans No. 8 will be W=88 pcs/min with rotation rate of the driven wheel n_{opt1}=30.6 r/min.

From the regression Equation (6) a two-dimensional section of resonance surface can be made. It characterizes dependence of cleaning quality on temperature of washing solution and its concentration with a fixed value of optimal rotation rate of the driven wheel n = 30.6 r/min (Figure 5).

$$Y=4.969+1.501\cdot x_1+14.378\cdot x_2-0.015\cdot x_1\cdot x_3-0.139\cdot x_2\cdot x_3-1.012\cdot x_2^2+0.038\cdot x_3^2. \quad (6)$$

From the analysis of dependence (Figure 5), it is seen that necessary cleaning quality with optimal rotation rate of the driven wheel can be reached due to different combinations of temperature (73.5…85 °C) and concentration of washing solution (1.73…3 g/l). Therefore, increase of requirements to the quality of cans surface leads to considerable expenditure of washing solution per liter of water and to increase of washing solution temperature. Moreover, concentration in the area of good cleaning increases quite intensively. That is why absolute purity of cans surface is inappropriate from the point of view of economy of can goods production.

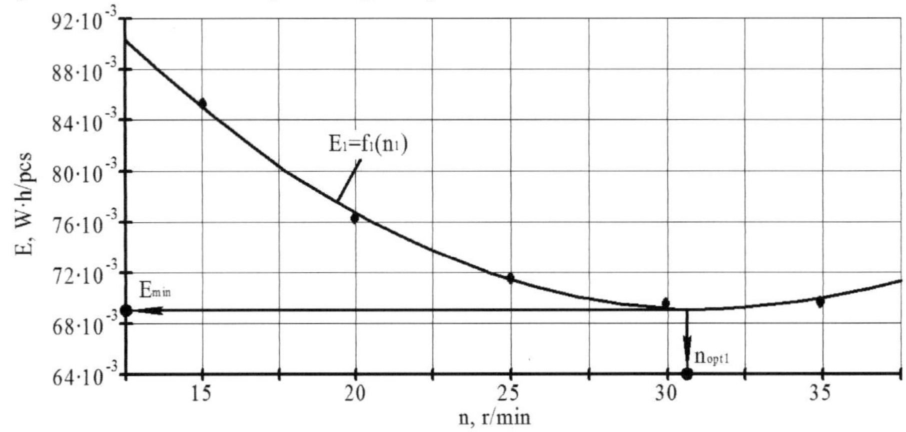

Figure 4. Dependence of energy output on rotation rate of the driven wheel.

With increase of rotation rate of the driven wheel by 14.38% (from 30.6 to 35 r/min) energy output increases only by 1.92% (from $68.9765\cdot10^{-3}$ to $70.3\cdot10^{-3}$ W·h/pcs). Production capacity also increases by 13.64% (from 88 to 100 pcs/min) that is important for cans production.

Committing concentration of washing solution c = 1.73g/l that is minimal due to optimal rotation rate, we make a two-dimensional section of resonance surface from the regression Equation (6) that characterizes dependence of cleaning quality on temperature of washing solution and rotation rate of the driven wheel (Figure 6).

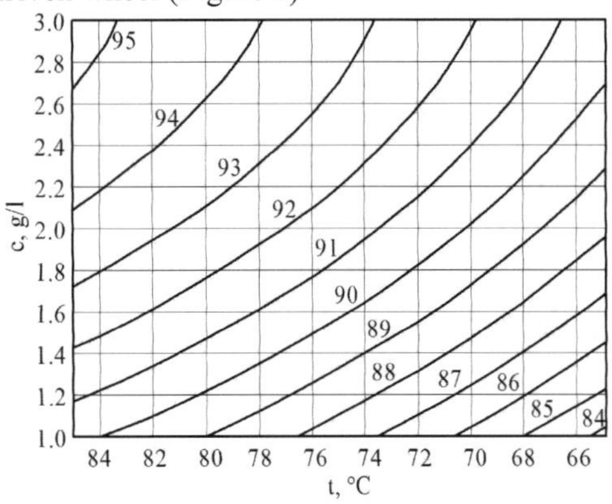

Figure 5. A two-dimensional section of resonance surface that characterizes dependence of cleaning quality (Y) on factors x_1 (temperature of washing solution, t) and x_2 (concentration of washing solution) with a fixed value of factor x_3 (n=30.6 r/min).

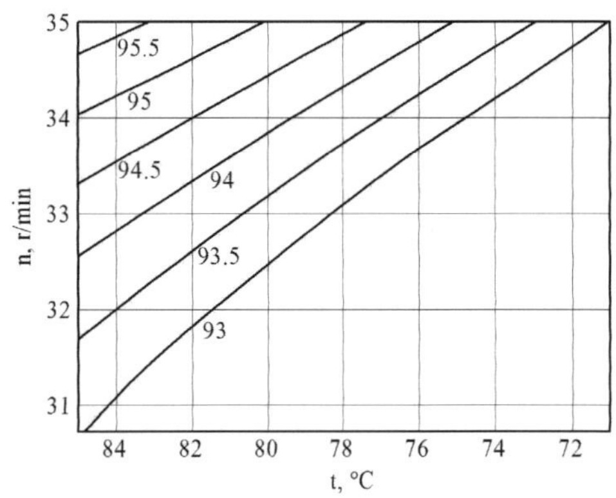

Figure 6. A two-dimensional section of resonance surface that characterizes dependence of cleaning quality (Y) on factors x_1 (temperature t of washing solution) and x_3 (rotation rate of the driven wheel, n_1) with a fixed value of concentration x_3 (c=1.73 g/l).

From analysis of dependence (Figure 6) it is seen that cleaning quality can increase to $K \geq 93$ % with concentration of solution $c = 1.73$ g/l, temperature from 72 °C to 85 °C and with rotation rate of the driven wheel from 30.6 r/min to 35 r/min.

4. Summary

According to the results of research, the following optimal parameters are defined: necessary cleaning quality with optimal rotation rate of the driven wheel can be reached with temperature (74...85°C) and concentration of washing solution of 1.7...3 g/l.

If it is necessary to increase production capacity from 88 to 100 cans per minute, it is important to increase rotation rate of the driven wheel from 31 to 35 r/min. Concentration of solution should be minimal due to optimal rotation rate and constitute $c = 1.7$ g/l. Energy output also increases by 2% that is not significant increase in comparison with increase of machine production capacity. Quality of cans surface cleaning will be K ≥ 93% with temperature from 72 to 85 °C.

References

[1] Rauf S *et al* 2016 *Procedia Comput Sci* **83** 902–906
[2] Abdel-Fatah M A *et al* 2017 *Ain Shams Engineering Journal* **8(3)** 305–310
[3] Wahad H S *et al* 2017 *IOP Conf. Ser.: Mater. Sci. Eng.* **174** 012021 doi: 10.1088/1757-899X/174/1/012021
[4] Jain S V *and* Patel R N 2014 *Renew Sust Energ Rev* **30** 841–868
[5] Akikur R K *et al* 2013 *Renew Sust Energ Rev* **27** 738-752
[6] Plappally A K *and* Lienhard V J H 2012 *Renew Sustain Energy Rev* **16(7)** 4818–4848
[7] Tadesse M G *et al* 2017 *Smart Mater Struct* **26** 065016

IOP Conf. Series: Materials Science and Engineering **457** (2018) 012013 doi:10.1088/1757-899X/457/1/012013

Research into spontaneous activity of myocardial cells under normal and pathological conditions using the hardware and software complex based on nanosensors

M M Yuzhakov[1], M L Ivanov[1], N V Turushev[1], I V Maksimov[2], S V Mazikov[1], R E Kodermyatov[1], G Wenjia[1], I A Zimin[2]

[1]R&D Laboratory for Medical Engineering, National Research Tomsk Polytechnic University, Tomsk, Russia
[2]Cardiology Research Institute, Tomsk National Research Medical Center of the Russian Academy of Sciences, Tomsk, Russia
E-mail: libra@tpu.ru

Abstract. Cardiovascular diseases are the leading cause of death in the world. The paper focuses on the capability of a nanosensor-based hardware and software complex (HSC) developed at Tomsk Polytechnic University to measure the activity of myocardial cells from the surface of the human body. A comparative study of nanosensors used in the HSC and conventional AgCl electrodes by FIAB Spa (Florence, Italy) was carried out. It is shown that the value of electromagnetic interference in conventional electrodes is several times higher compared to nanosensors. ECG was recorded using the developed HSC in order to show the possibility to control the activity of myocardial cells.

1. Introduction

One of the most common pathologies encountered in medicine is the heart pathology leading to death. Out of the total deaths worldwide, diseases of cardiovascular system occupy leading positions [1]. World Health Organization statistics 2015 showed that 17,866,560 people died from diseases of the cardiovascular system. Among all the causes of death from cardiovascular diseases, the phenomenon of sudden cardiac death (SCD) is prevailing. SCD is an unexpected death due to cardiac causes. The phenomenon occurs within a short period of time and manifests itself in immediate and severe symptoms that hinder effective treatment and prophylaxis [2]. In this regard, the study of spontaneous activity of myocardial cells is an important area of research. The analysis of the sources shows that the methods of SCD prevention has not changed considerably over the last 35 years [3–6].

Given the importance of this problem, a lot of attention has recently been paid to finding risk markers for SCD. These risk markers include demographic indicators, low left ventricular ejection fraction, decreased heart rate variability, ventricular rhythm disturbances: stable ventricular tachycardia, frequent ventricular extrasystole (> 10 per hour), short paroxysms of ventricular tachycardia, late ventricular potentials [6], microvolt T-wave alternans, high-resolution electrocardiography indicators such as the filtered QRS interval (fQRS) and late ventricular potentials, variance and daily dynamics of the QT interval, variability, heart rate turbulence, and baroreflex sensitivity [3]. At the same time, different literature sources give different data on the prognostic significance of these markers. The significance of some markers is disputable [7].

Due to the fact that the positive predictive value of each risk marker for SCD is not high, a combination of features is proposed to be used [6]. Unfortunately, at the moment there is no generally accepted algorithm for examining patients, which makes it possible to assess the risk of SCD.

This suggests that it is necessary to look for new features and criteria that will allow us to identify high-risk groups of SCD.

2. Methods and Materials

Currently, efforts of many scientists and specialists are aimed at solving problems of automatic analysis and recognition of cardiovascular pathologies. For this, various methods and algorithms for processing and analyzing ECG are used. This is evident from a large number of publications and scientific papers devoted to this problem [8–11]. A number of methods have been developed. The most common methods for identifying R-peaks are based on the analysis of the first derivative of ECG [9]. The methods are aimed to provide the user with the most accurate data of measurements taken after their processing. This is extremely important in view of the fact that any transformation of a measuring signal implies its distortion depending on the method chosen. Filtering significantly distorts the signal [12].

Tomsk Polytechnic University in collaboration with Tomsk Cardiology Research Institute (TCRI) have developed a high-resolution nanosensor-based hardware and software complex (HSC) for real-time recording of micropotentials of the heart without filtering and averaging. HSC comprises nanosensors developed at Tomsk Polytechnic University [13]. Nanosensors are of high stability, low noise and noise immunity of unpolarized electrodes. Using these nanosensors, ECG can be recorded without the use of filtering units in the frequency range from 0 to 10,000 Hz. In addition to standard metrological measures, a series of experiments was conducted to confirm a high quality of recording of physiological signals from the body surface performed using nanosensors. For this purpose, AgCl electrodes by "FIAB Spa (Florence, Italy)" were used as a reference for comparison. The measurements were carried out under similar environmental conditions. The same patient was taken for measuring, and the study method was also unchanged.

The preliminary studies to verify the quality of HCS were carried out in volunteers and patients of TCRI. Each volunteer signed the informed consent form for clinical trials (clinical trials were approved by the local ethical committee for biomedical ethics, TCRI, protocol No. 147 dated June 28, 2016). The volunteers were provided with sufficient information on the procedures of examination and treatment, according to the internal regulations on the procedure for hospitalization in TCRI. The results of the study maintain confidentiality of the patients and volunteers surveyed.

3. Results and considerations

The results of comparison of nanosensors and conventional AgCl electrodes by FIAB Spa (Florence, Italy) is presented in Figure 1. These results show that AgCl electrodes exhibit significant interference of 50 Hz and its harmonics under the same measurement conditions. Conventional electrodes can only be used with filters (including power-line filters). The interference is usually random, both in terms of the level and the width of the spectrum in a 50 Hz region. A power-line filter may significantly distort micropotentials of the heart [12], and dynamic observation will not provide reliable results.

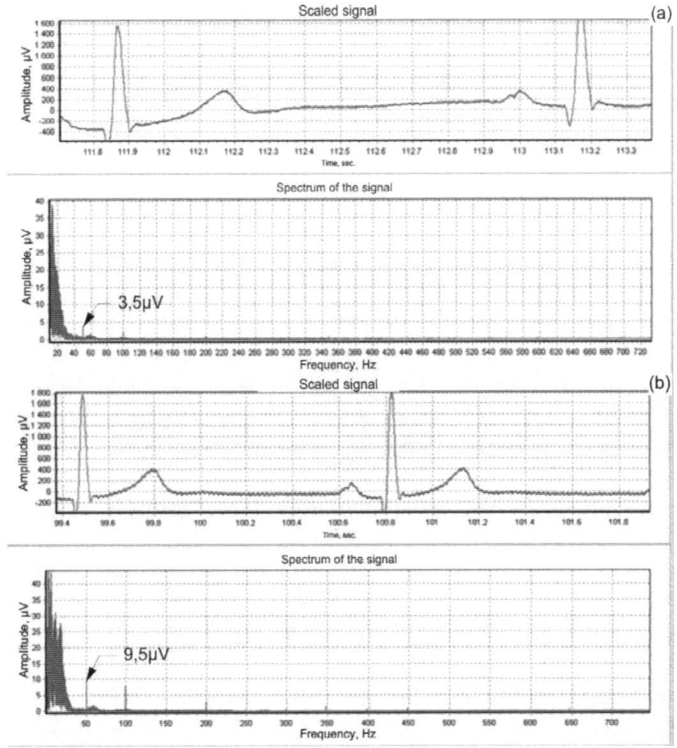

Figure 1. Comparison of electromagnetic pickups of (*a*) the nanosensor (sensitivity 200 µV/div) and (*b*) the conventional AgCl electrodes by FIAB Spa (Florence, Italy) (sensitivity 200 µV/ div).

Figure 2*a-c* presents patient electrocardiogram # 178 records in the frequency range from 0 to 10,000 Hz. Registration was carried out in the medical ward away from sources of electromagnetic interference (including medical equipment).

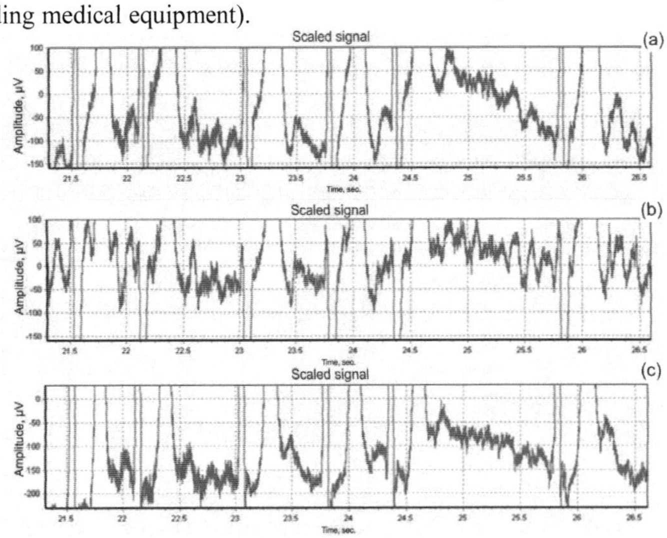

Figure 2. High-resolution ECG recording of patient # 178, frequency range from 0 to 10,000 Hz. (*a*) - 1 lead; (*b*) - 2 lead; (*c*) - 3 lead.

IOP Conf. Series: Materials Science and Engineering **457** (2018) 012013 doi:10.1088/1757-899X/457/1/012013

4. Conclusion

Figure 1 shows that nanosensors have a higher sensitivity and improved signal-to-noise ratio. Noise in the ECG recorded using nanosensors is less significant than that recorded by conventional electrodes. ECG recorded far from the sources of interference does not contain electromagnetic interference in the range from 0 to 10,000 Hz. As can be seen from Figure 2, ECG recording performed by the HSC (including spontaneous activity of myocardial cells) does not require any filtering and averaging.

References

[1] Mahanani W R 2013 Projections of mortality and causes of death, 2015 and 2030 *World Heal. Organ*
[2] Zipes D P and Wellens H J J 1998 Sudden Cardiac Death *Circulation* **98** 2334–51
[3] Golukhova E Z *et al* 2017 *Kardiologiia* **57** 73–81
[4] Spirito P and Maron B J 1990 *J. Am. Coll. Cardiol.* **15** 1521–1526
[5] Maron B J *et al* 1982 *Circulation* **65** 1388–1394
[6] Priori S G *et al* 2001 *Eur. Heart J.* **22** 1374–1450
[7] Malik M 2001 *Risk of Arrhythmia and Sudden Death* (London: BMJ)
[8] Elgendi M 2013 *PLoS One* **8** e73557
[9] Djohan A *et al* 1995 *Proceedings of 17th International Conference of the Engineering in Medicine and Biology Society* **1** 167–168
[10] Martis R J *et al* 2013 *Biomed. Signal Process. Control* **8** 437–448
[11] Rybalka S *et al* 2018 *MATEC Web Conf.* **155** 1008
[12] Avdeeva D K *et al* 2014 *Biol. Med.* **6** BM-025
[13] Mishchenko K V *et al* 2016 *Eur. J. Nanomedicine* **8** 195-202

IOP Conf. Series: Materials Science and Engineering **457** (2018) 012014 doi:10.1088/1757-899X/457/1/012014

Technique for monitoring the indicators of the total contact spot of spur gears of the truck's differential gear

D T Safarov, A G Kondrashov

Associate Professor, Naberezhnochelninsky Institute (branch) of Kazan (Volga Region) Federal University, Naberezhnye Chelny, Russia
E-mail: safarov-dt@mail.ru

Abstract. The article contains the main provisions of the technique, which allows to control the area of the total contact patch and its position relative to the tooth edges. Graphical schemes for finding indicators are given. An example of the calculation of the indicators established by the methodology, as well as the results of its application for assessing the quality of the setup performance, is considered on the example of processing a differential satellite in the operation of circular gearing.

1. Introduction

Straight bevel gears are part of various differential gears to ensure the independence of the rotation of the driving wheels of the car.

The accuracy of the processing of the teeth of conical wheels largely determines their resource and noise level in the transmission. One of the main indicators determining the reliability of the conical transmission in the differential is the total contact spot. Inconsistencies in the location of the contact spot in the conical gear lead to serious consequences during the operation of the vehicle.

For example, the local output of the contact patch to the spherical end face of the satellite leads to the appearance of a significant moment directed along the axis of the satellite. Chamfering at the ends of the teeth can not fully compensate for the exit of the contact patch on the edge [1, 2]. As a consequence, the pressure in the area of the hole close to the flat end of the satellite increases. There is an extrusion of the oil film, heating and wear of the antifriction coating and adhesion of the satellite to the reference axis. After that there is jamming of the gear drive, gear transmission, breakage of the teeth, destruction of the mated parts and failure of the differential.

2. Statement of the problem

The normalization of the total contact spot is carried out according to the requirements of the standards. So the requirements of the standard [3] for bevel gears, the relative dimensions of the total contact patch in percent are determined: the length of the tooth is the ratio of the distance between the extreme points of the traces of adherence to the length of the tooth; on the height of the tooth - the ratio of the average height of the traces of adherence to the average tooth height of the corresponding active lateral surface (Figure 1).

An estimate of the contact spot size is performed according to the standardized procedure given in [5]. The standard defines the concept of an instantaneous contact spot as one of the indicators of the accuracy of tooth contact. An instantaneous contact spot means "a part of the active side surface of the gear wheel tooth on which traces of its attachment to the gear teeth coated with the dye are located after turning the wheel of the assembled gear to full turn with slight braking ensuring continuous contact of the teeth of both gears" [5].

Relative dimensions of the contact spot in percent:

- by the length of the tooth - the ratio of the distance between the extreme points of the traces of adherence to the length of the tooth ($\frac{a}{b} \times 100$);
- by the tooth height - the ratio of the average height of the traces of adherence to the average tooth height of the corresponding active side surface ($\frac{h}{h_a} \times 100$).

Currently, many machine-building enterprises, as well as assembly enterprises, assess the shape and dimensions of the total contact spot visually. In this form, the method is used for the initial assessment of the correspondence of a single tooth of the wheel toothed wheel, as well as the adjustments of the machine nodes to achieve the location of the spot, was approximately at the center of the lateral surface of the tooth. In rare cases, the dimensions of the contact spots along the length and height are determined for one or two teeth of the wheel by measurements using universal measuring instruments.

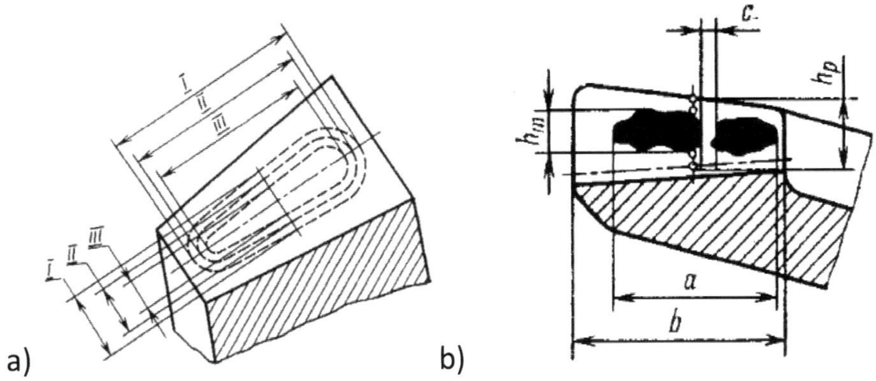

Figure 1. *a*) Normalizing the size of the contact patch in a bevel gear. I is the largest limiting size of the contact spot; II - nominal size of the contact spot; III - the smallest limiting size of the contact spot [1]; *b*) the standard scheme for estimating the dimensions of the total contact spot [4].

It should be noted that the standard method is based on the assumption of the location of the contact spot near the calculated point of the gear train. Its calculated data can not be used to assess the compliance of the total contact spot with significant deviations from the calculated point of transmission, for example, in the adjustment processes or the deviations that occur when the bevel gears are chiseled.

3. Theoretical part

Because the tapered differential gears of the truck are not adjustable, it is not possible to change their axial position during installation during the assembly of the differential and thus affect the position of the total contact spot. It is important to obtain a contact patch not only approximately in the central part of the gear tooth, but to ensure the stability of its location along all the teeth of the ring gear after the operation of the gear.

The developed procedure for normalizing the total contact patch takes into account the requirements of the standard (allows quantifying the dimensions of the contact patch along the axis and tooth height), and additionally normalizes its relative position on the side surface of the gear wheel (Figure 2). Let us consider the main provisions of the methodology for finding separate indicators of the quality of the total contact spot.

Based on the data of the conical transmission tests under load, it is established that the optimal form of the total contact spot is an ellipse located at a distance of two to four mm from the flat end of the conical transmission satellite (Figure 2*a*). The minimum width of the contact patch in its middle part (or the small axis of the ellipse) should be ½ the active height of the tooth. Along the axis of the tooth, the total contact spot (or major axis of the ellipse) should occupy from ½ to ⅔ its length along the generatrix of the fission cone. These data are taken into account when determining the minimum and maximum allowable form of the total contact spot (Figure 2*b*).

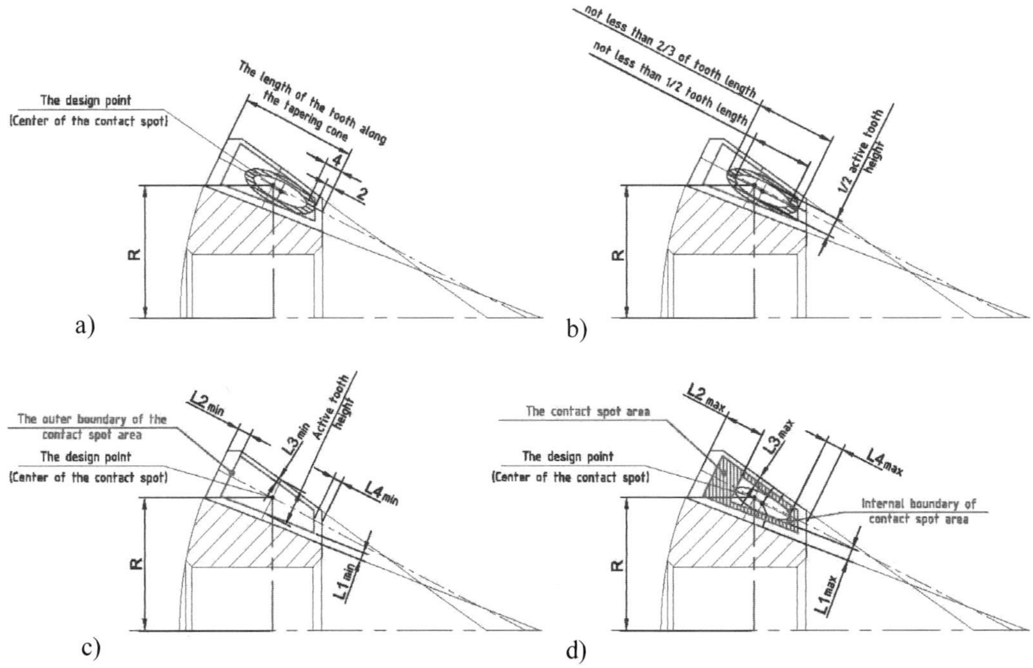

Figure 2. Normalization of the shape and relative location of the total contact patch *a*) nominal location, *b*) permissible sizes of the total contact spot, *c*) the maximum permissible detection limits, *d*) the minimum permissible detection limits.

The normalization of the boundaries of the location of the total contact spot is to establish the minimum and maximum boundaries for finding the total contact spot (Figure 2*c, d*). The boundaries of the zone are located in the form of lines that equidistantly repeat the contours of the tooth at a given distance from the edges of the tooth. Depending on the gear module and the magnitude of the moment transmitted by the gear, it is 1–3 mm.

As a result, the form of the actual contact patch should be between ellipses and should not go beyond the inner and outer boundaries of the location area. The obtained generalized values of the indicators allow us to calculate the limiting values of the boundaries of the tolerance fields and are suitable for calculating the values of the gears of any conical spur gear.

4. Practical implementation

The contact spots are formed as a result of the rolling of the teeth on a special gear-rolling machine (Figure 3).

IOP Conf. Series: Materials Science and Engineering **457** (2018) 012014 doi:10.1088/1757-899X/457/1/012014

a) 

b) 

Figure 3. Formation of the total contact spot: *a*) the rolling mill, *b*) the result of rolling the teeth of the gears.

Quantitative determination of indicators is possible only on the basis of photographing the lateral surfaces of the gear teeth with further processing of the image (Figure 4*a*). The image processing consists of the outline of the tooth contours and contact spots, which resulted from abrasion of the special pigment. To estimate the dimensions of the total contact patch, schemes have been developed for finding the parameters of the total contact spot in accordance with the requirements of the provisions of the technique of simultaneous normalization and the shape and location of the total contact spot (Figure 4*a*, *b*).

a) 

b) 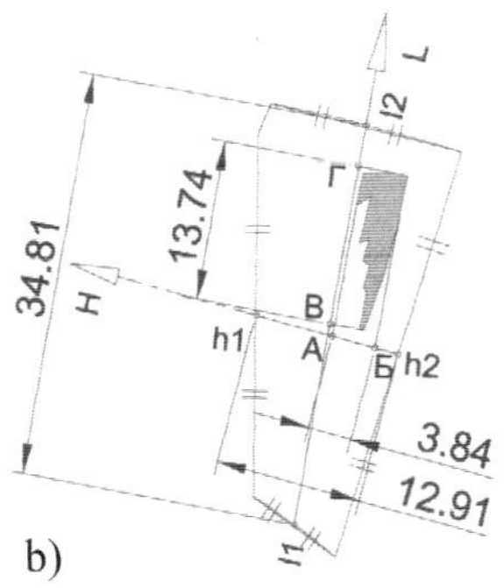

c) 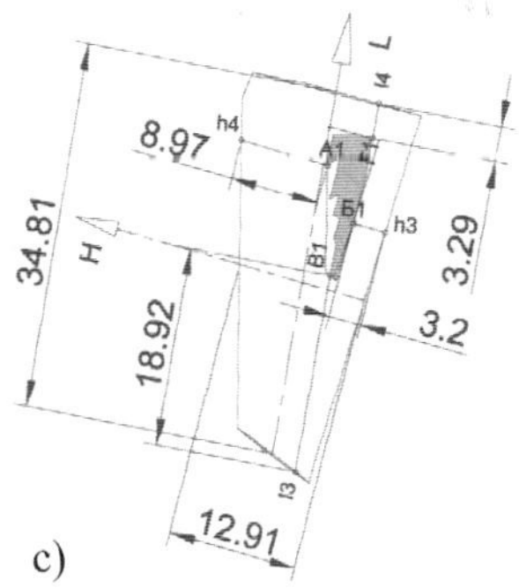

Figure 4. Graphical schemes for finding the parameters of the total contact spot: *a*) photograph of the side surface of the gear wheel with the contours of the active zone of the lateral surface and the contact spot, *b*) the scheme for finding the dimensions of the total contact patch, *c*) the scheme for finding the parameters of the location of the total contact spot.

Calculation of the area and relative location of the total contact spot for the calculation schemes shown in Figure 4 is performed by the following formulas:

1. The size of the total contact patch along the length of the tooth.

IOP Conf. Series: Materials Science and Engineering **457** (2018) 012014 doi:10.1088/1757-899X/457/1/012014

$$\%L = \frac{a}{b} \times 100 = \frac{13.74}{34.76} \times 100 = 39.53\ \%$$

2. The size of the total contact patch along the height of the tooth.

$$\%H = \frac{h}{h_a} \times 100 = \frac{3.82}{12.91} \times 100 = 29.59\ \%$$

3. Calculation of the linear dimensions of the location of the total contact patch along the length of the tooth:

$$L2 = \frac{L2}{b} = \frac{3.29}{34.81} * 26.44 = 2.45\ mm$$

$$L4 = \frac{L2}{b} = \frac{18.92}{34.81} * 26.44 = 14.37\ mm$$

4. Calculation of the linear dimensions of the location of the total contact spot over tooth height:

$$L1 = \frac{L2}{h_a} = \frac{2.91}{12.91} * 8.18 = 1.84\ mm$$

$$L3 = \frac{L2}{h_a} = \frac{8.97}{12.91} * 8.18 = 5.68\ mm$$

The results of the simultaneous calculation of the area of the contact patch serve for the construction of control charts for controlling the progress of the adjustment of gear machines with the change of circular broaches.

5. Conclusions and results

An example of control charts for %L for rolling clockwise and counterclockwise is shown in Figure 5. The control map field indicates the outlets of the graph points beyond the limits of the tolerance zone USL and LSL (the red line on the control field is shown), which indicates the presence of special reasons. In addition, the estimation of the intrinsic variability of the parameter (on the field of the control map indicated by dashed lines) is performed. In the process of adjusting by radial, axial and angular displacements, broaches seek to eliminate on the charts the presence of special causes and hit the control boundaries of the graphs for the remaining set of indicators - %H , L2 , L4 , L1 , L3 .

In addition to ensuring control over the implementation of adjustment, the developed methodology has shown its effectiveness and effectiveness in the certification audit of process equipment in the workplace of circular tooth gearing of differential gears. The approach presented in the methodology can be successfully used for the purpose of improving the quality of products in a manner similar to those described in [6–8].

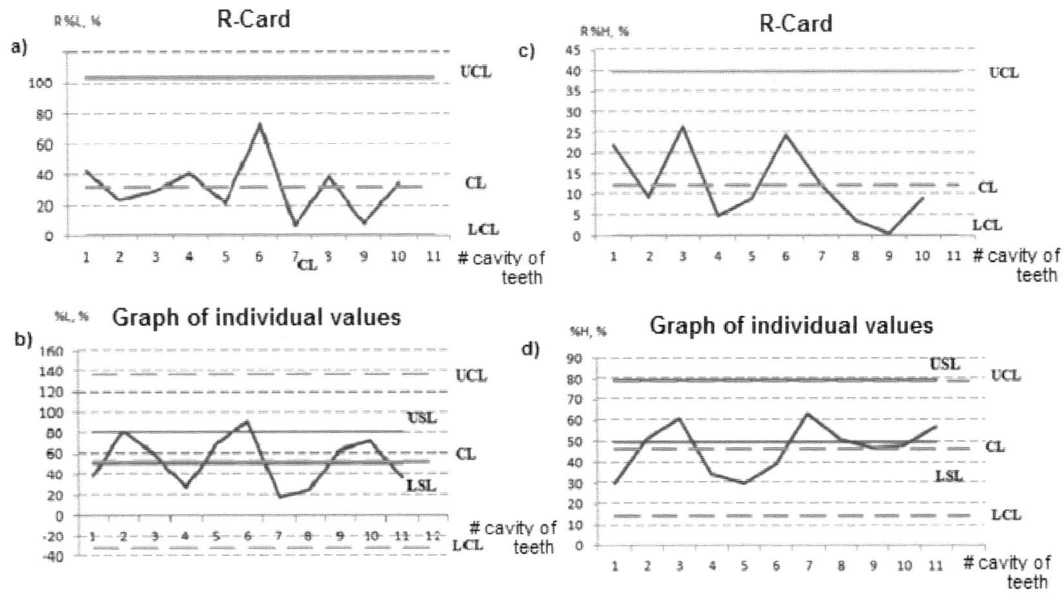

Figure 5. Example of control charts of individual values and sliding ranges of the% L and% H indices on the side surfaces of the satellites crown during the adjustment stage. *a, b*) when the satellite is turned in a counterclockwise direction, *c, d*) while rotating clockwise.

Thus, the quantitative data characterizing the area and the relative location of the contact spot make it possible:

- The use of control charts to eliminate special causes in gear tooth gear cutting, the preparation of a new product and the adjustment of the machine.
- Calculate the indices of stability and tuning of the gear cutting processes for the purpose of attestation of the technological process.
- To provide more effective control over the adjustment of the machine already in the adjusted process of gear cutting.

References

[1] Chemborisov N A, Kondrashov A G 2008 *Russian Engineering Research* **28**(8) 809–811 doi: 10.3103/S1068798X08080182

[2] Kondrashov A G , Safarov D T, Faskhutdinov A I , Davletshina G K 2017 *Russian Engineering Research* **37**(9) 812–813 doi: 10.3103/S1068798X17090143

[3] GOST 1758-81 Basic norms of interchangeability. Bevel and hypoid gears. Tolerances. Moscow, Izdatelstvo Standartov (in Russian)

[4] GOST 1643-81 Basic requirements for interchangeability. Cylindrical gearings. Tolerances. Moscow, Izdatelstvo Standartov (in Russian)

[5] GOST 9368-81 Basic norms of interchangeability. Fine-pitch bevel gears. Tolerances. Moscow, Izdatelstvo Standartov (in Russian)

[6] Kasjanov S V, Kondrashov A G, Safarov D T 2017 *Procedia Engineering* 206 1508–1514 doi: 10.1016/j.proeng.2017.10.670

[7] Safarov D T, Fedorova K A and Ilyasova A I 2016 *IOP Conf. Series: Materials Science and Engineering* **134** 012036 doi: 10.1088/1757-899X/134/1/012036

[8] Safarov D T, Kondrashov A G 2018 *Journal of Physics: Conference Series* **1050** 012075 doi:10.1088/1742-6596/1050/1/012075

IOP Conf. Series: Materials Science and Engineering **457** (2018) 012015 doi:10.1088/1757-899X/457/1/012015

Improving the functionality of the semiconductor matrix receiver of optical radiation

A A Shnyakin, E Ph Pevtsov, T A Demenkova

Associate professor, MIREA - Russian Technological University, Moscow, Russia
E-mail: pevtsov@mirea.ru

Abstract. The work is devoted to improving the functionality of the semiconductor integrated matrix receiver of visible range radiation with a capacitor multichannel analog-to-digital Converter on a chip. Reading is carried out by a column capacitor of sequential approximation with a digital-analog converter based on switched capacitors DAC with charge summation. In the matrix receiver, the occupied area and the dissipated power of the semiconductor chip are significantly reduced with the simple implementation of capacitor structures by the technology of semiconductor micro-devices manufacturing.

1. Introduction

Analysis of papers on metal-oxide-semiconductor (CMOS) image sensors reveals current developmental trends aimed to combine functions of converting radiation into electrical signals, digitizing them and digital processing in a single chip or package [1–3].

There are three main approaches as regards placing an ADC in a readout integrated circuit chip, that is, a chip-level ADC, a column-level ADC and pixel-level ADC. Column-level ADCs may have complex architecture and combine various ways of analog-to-digital conversion since its layout implementation is only one-dimension limited and is not limited in other one. The speed of such ADCs is comparable to the rows' readout rate of a readout IC which positively affects power consumption. Thus, using a column-level ADC implies balancing such factors as speed, conversion accuracy, power consumption and chip area [4, 5].

The main goal of this work was to prototype ADC IP-core intended for reuse in different families of CMOS image sensors. The circuitry and layout of resulting design solutions are determined by the following main factors: optimization of production costs and detectors' characteristics; minimization of the area occupied by ADC units; minimization of power consumption upon reaching at least twelve bits of conversion for at least 1.0 MHz clock frequency and 77000 op/sec working speed.

2. Condenser multi-channel ADC

Widely used optimal solution for the ADC in the composition of the semiconductor structures for complementary CMOS receivers is a diagram of a successive approximation ADC based on a digital-to-analog converter Binary-Weighted capacitor DAC, capacitance which is a multiple of a power of two. This is due to the relative ease of implementation of capacitor structures in the framework of standard technological processes of modern VLSI. The main disadvantage of this solution is the area of capacitor structures, increasing with the growth of the conversion bit width. Accordingly, the currents required for charging and discharging them and the dissipated power increase significantly.

IOP Conf. Series: Materials Science and Engineering **457** (2018) 012015 doi:10.1088/1757-899X/457/1/012015

The peculiarity in the proposed version of the ADC in which an attempt is made to correct this drawback is that where the DAC circuit is made with three subunits of weighing capacitors, each subunit is supplied with different reference voltages, the values of which correspond to a certain range of the conversion bit width [6].

The composition of the ADC includes: voltage comparator, parallel DAC on switchable capacitors with charge summation and digital control circuit (successive approximation register). The main elements of the DAC are: a matrix of weighing capacitors with switchable keys and an operational amplifier with a capacitor and a key in the negative feedback circuit.

The basis for the design of the topology of the complex functional ADC module was the PDK (process design kit), the base gate width of the transistor cell of which was 0.54 μm. Multiple ratios between the capacitance values of the capacitors are provided by multiplying the cell of the dual capacitor, and the subsequent parallel Association of the multiplicated fragments of the topology in accordance with the required capacitance value.

Simulation of the ADC electrical circuit Assembly was carried out using models of devices included in the customer-provided library of parameterized cells, CMOS-oriented technology with design standards of 0.35 μm.

The simulation with the extracted parasitic parameters of the topology is obtained, that the conversion efficiency is up to 10 bits without correction. When adjusting the DAC by varying the reference voltage and the mixing voltage, the ADC efficiency can be increased to 11-12 bits.

The design solutions and GDS-files of the topology of successive approximation column ADCs intended for placement on a chip with CMOS photomatrix are obtained.

3. Analyzing Test Chip Characteristics

In order to analyze test chips for the ADC IP-core and CMOS image sensors, a hardware and software was developed that allows recording automatically dynamic and static characteristics of the tested devices and their components. Structure of the test equipment is shown in Figure 1.

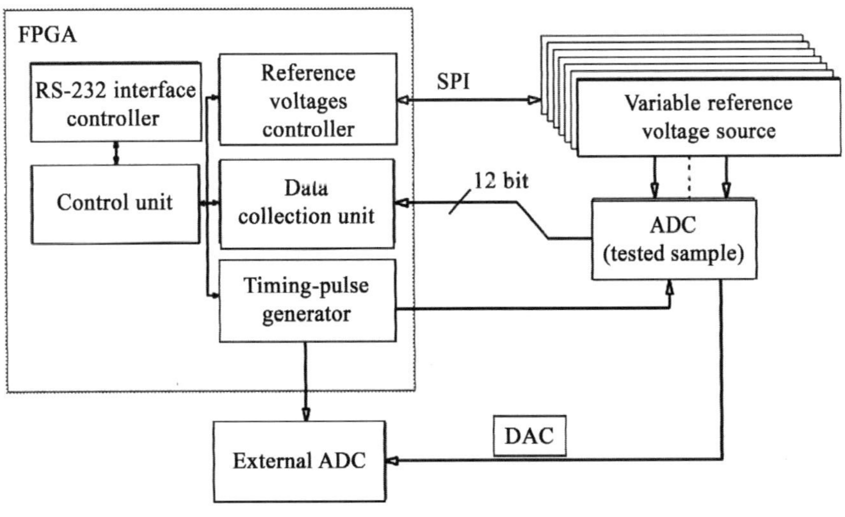

Figure 1. Schematic structure of the test stand.

The main stand element is the process controller, made as Spartan 3E family FPGA chip firmware. The main functions of the controller are the following: 1) to control the DAC; 2) to collect data from the tested ADC; 3) to form a clock cycle and output signals to synchronize the external ADC which is used to measure parameters of the device under test. To describe the controller behavior, the VHDL language and the standard XILINX libraries were used. Software set for testing the ADC provides adjustment of reference voltages, creating a description of the clock diagram and its loading into the

IOP Conf. Series: Materials Science and Engineering **457** (2018) 012015 doi:10.1088/1757-899X/457/1/012015

generator's memory, programming the FPGA, data collection and display. Software includes: a program for setting modes and measuring parameters, a clock pattern editor, tools for controlling operation of the InSys ADM216x100M external ADC and displaying the obtained data. Device resolution can be programmed for 20 bits with output voltage accuracy up to 0.003 % reached within 15 ms.

Analyzing test chips of the ADC IP-core was carried out as follows. First of all, operability of each sample was confirmed by sequentially switching each of the three DAC units as a Most Significant Bit Block, meanwhile the remaining units were disabled. When supplying alternately each unit under test with 1.5 V as the reference voltage and 2.5 V as the common reference voltage, the ADC input voltage from the test stand DAC varying within the 2.5316 V to 3.43785 V range in 0.0156 V increments, the results for each tested unit at the DAC reference output point of the tested ADC were the same within the measurement error. Data from the ADC digital outputs were also identical. Operation of the ADC as a whole was tested with normal switching of all three capacitor units and the calculated from expressions reference voltages for the input signal varying all over the range of input voltages.

Studying linearity characteristics of the DAC circuit showed significant deviation from the required and calculated values (Figure 2) due to stray capacitances between digital control lines and the common bus of capacitor structures in each of the twelve digital units of switch control, storage and output which was ignored during the initial design.

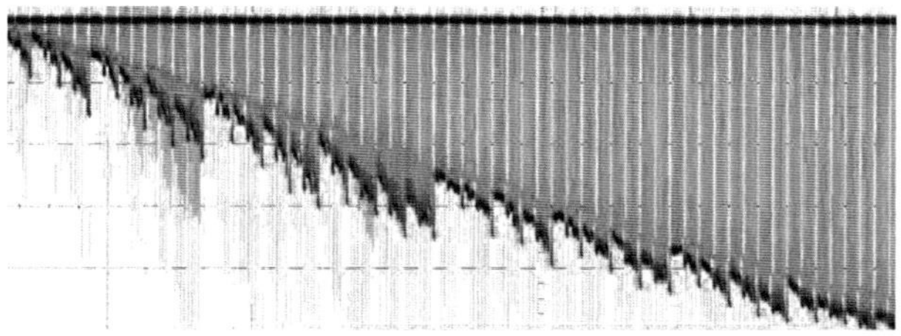

Figure 2. Fragment of the DAC transfer characteristic.

A layout chart fragment of a digital unit with marked places where stray capacitive couplings were formed is shown in Figure 3.

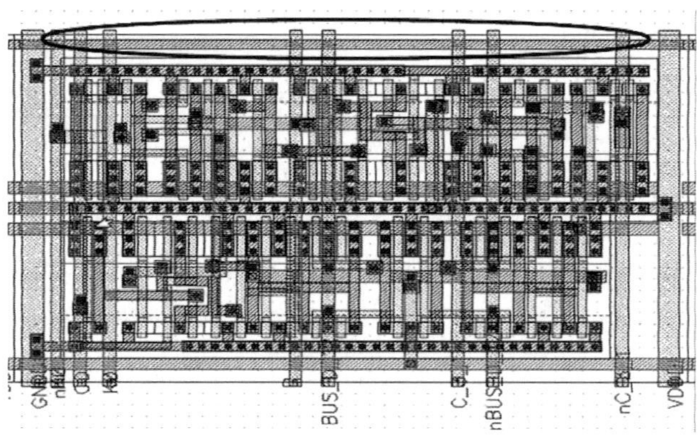

Figure 3. Layout chart fragment of a digital unit of switch control, storage and output.

The DAC circuit linearity characteristics are negatively affected by capacitive parasitic cross-talks of switch control digital buses and the common bus of capacitor structures which was confirmed by the results of measurements upon specifying certain conditions. If reference voltages are applied to the ADC inputs, the DAC output voltage must remain unchanged for any combination of control signals.

IOP Conf. Series: Materials Science and Engineering **457** (2018) 012015 doi:10.1088/1757-899X/457/1/012015

As a result, it was decided to eliminate capacitive parasitic cross-talks by grounded shielding metallization layer between digital control buses and the common bus of capacitor structures. However, experiments of the test chips' sample quantity of the layout adjusted ADC IP-core revealed complete failure for all the investigated samples containing a single column of the ADC IP-core. By comparing layout by means of the DRC and LVS CAD (Figure 4*a*), and the test chip layout by means of optical and electron microscopy (Figure 4*b*), the cause of the failures was found to be defects which caused short-circuiting of the external control enable bus and the supply voltage bus.

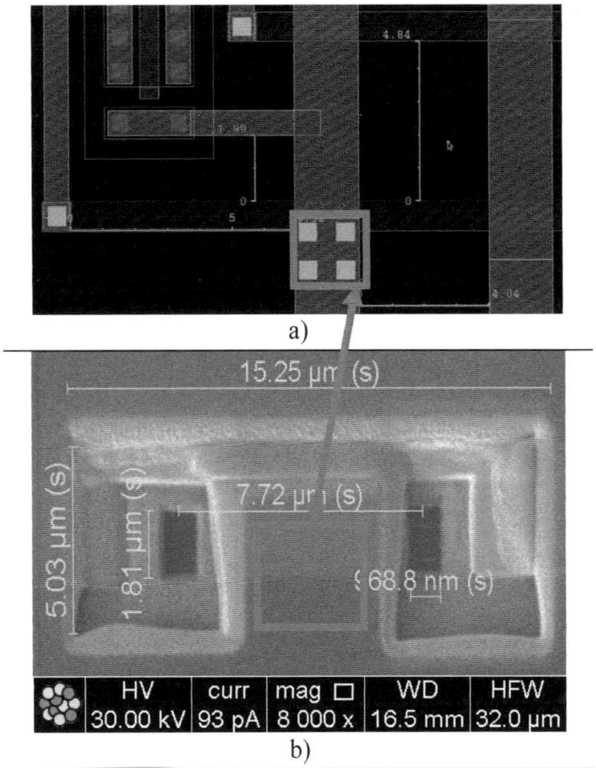

Figure 4. Localization and elimination of circuit closure.

To eliminate the identified defects, a method to correct test chip layout by means of ion-beam etching and ion-stimulated deposition based on focused ion beam (Figure 4*b*) was developed and applied in practice.

Optimization of ion beam formation allowed getting images of underlying metallization layers to adjust on a chip and localize short-circuiting of conduct buses without complete removal of a passivation layer. By means of ion-beam etching, all layers of inter- and intra-layer dielectrics were opened, and the short circuited section of the bus was separated. Then, by means of ion-stimulated deposition in a special installation with sharply focused ion beam (NanoLab complex with FIB), previously damaged integrity and conductivity of the control bus has been restored by a platinum layer. In this way, the technique aimed to restore working capacity of ADC chips was tested.

Analyzing chips of the second revision showed that the proposed change in the layout chart where influence of stray capacitive coupling was eliminated by introducing a grounded screen metallization layer, allowed to significantly reduce the DAC output voltage deviations (Figure 5) which as a result led to adjustment of data in the DAC transfer characteristic (Figure 6).

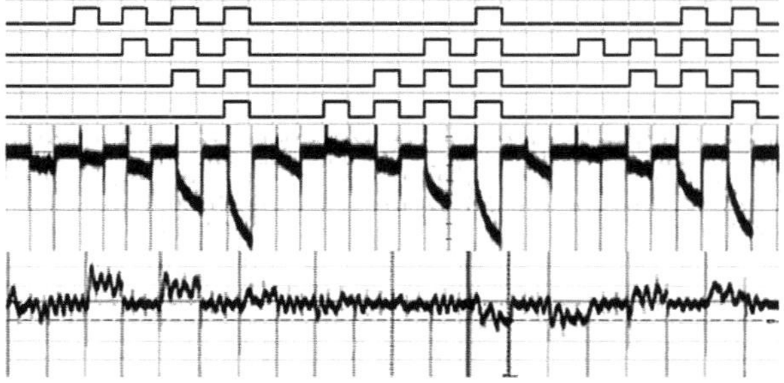

Figure 5. *a*) Flow diagram of control signals; *b*) DAC output voltage before adjustment; *c*) DAC output voltage after adjustment.

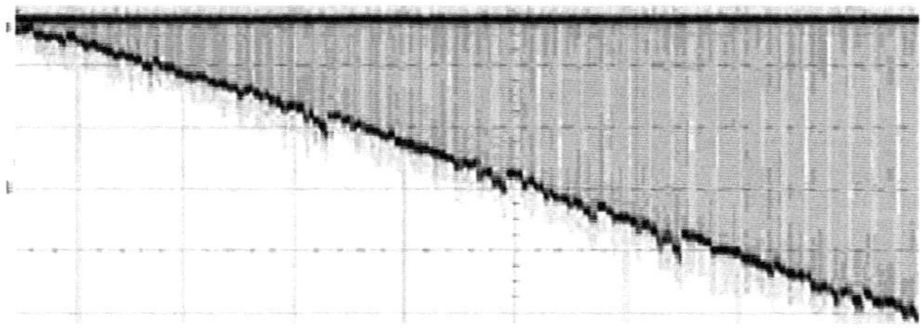

Figure 6. Fragment of the DAC transfer characteristic after adjustment.

4. Conclusion

The main characteristics of a capacitor multichannel analog-to-digital Converter in a semiconductor integrated circuit of a matrix radiation receiver are investigated. Circuit and layout solutions for a 12-bit column-level ADC were developed which allowed performing conversion in 12 cycles. When using the 0.5 μm CMOS technology level, the frame rate for the 1024×1024 pixel array is 75 Hz, while using the 0.18 μm technology level it will reach 1 kHz. The main design problems are related to strictly limited power consumption, limited ADC layout sizes (30 μm width for 15×15 μm^2 cells and dual-sided arrangement of ADCs) and the need to minimize stray capacitances that strongly affect conversion accuracy. To reduce the area of a capacitor divider, the ADC uses a novel three-stage circuit with three weighted reference voltage sources.

Acknowledgments

This work was supported by the Ministry of Education and Science of Russian Federation (project 8.5098.2017/8.9 basic part of the State task)

References

[1] Baker R J *et al* 1997 *CMOS Circuit Design, Layout, and Simulation*, Wiley-IEEE Press.
[2] Normanov D, Osipov D 2013 *Proceedings of the 12th WSEAS International Conference on Circuits, Systems, Electronics, Control & Signal Processing (CSECS '13)* Budapest. 90-94
[3] Pevtsov E Ph *et al* 2017 *IOP Conf. Ser.: Mater. Sci. Eng.* **168** 012095. doi: 10.1088/1757-899X/168/1/012095
[4] Ilan E *et al* 2013 *Proc. SPIE* **8659** A-1-0A-6. doi: 10.1117/12.2002845
[5] Gao L *et al* 2012 *Proc. SPIE* **8562** 10-1-10-6. doi: 10.1117/12.981856
[6] Borodin D V, Osipov Yu V 2017 *Television Technique series* 2 25–30 (in Russian)

IOP Conf. Series: Materials Science and Engineering **457** (2018) 012016 doi:10.1088/1757-899X/457/1/012016

Synthesis and Properties of Bi-HTSC synthesized on the basis of precursors obtained by melt quenching

D Uskenbaev[1], A Nogay[2], G Mukhamedrakhimova[3]

[1]Senior lecturer, Radio engineering, Electronics and Telecommunications department of S. Seifullin Kazakh Agrotechnical University, Astana, Kazakhstan
[2]Professor, Radio engineering, Electronics and Telecommunications department of S. Seifullin Kazakh Agrotechnical University, Astana, Kazakhstan
[3]Professor, Radio engineering, Electronics and Telecommunications department of L.N. Gumilyov Eurasian National University, Astana, Kazakhstan
E-mail: usdan@mail.ru

Abstract. The article presents the research results on the synthesis of HTSC ceramics using concentrated radiant radiation. It was established that all synthesized samples of the nominal composition $Bi_{1,7}Pb_{0,3}Sr_2Ca_{n-1}Cu_nO_y$ (n = 3, 4, 5), the main phase was the superconducting phase $Bi_{1,7}Pb_{0,3}Sr_2Ca_2Cu_3O_y$, which has a critical temperature of the transition to the superconducting condition of 107 K. Also in samples of nominal compositions $Bi_{1,7}Pb_{0,3}Sr_2Ca_{n-1}Cu_nO_y$ (n = 4, 5) showed traces of phases with higher T_c.

1. Introduction

After the discovery of high-temperature superconductivity (HTSC), various compositions (yttrium, bismuth, thallium, mercury, etc.) and methods for obtaining (superphase, melt, glass-crystalline, chemical, etc.) superconducting materials have been developed to increase the critical parameters (J_c) and the critical temperature (T_c). Among the above mentioned compositions, superconducting materials based on the Bi-Sr-Ca-Cu-O system are recognized as one of the most promising, since they are characterized by high critical characteristics, much less degradation, and consistency of the composition.

Analysis of the synthesis effect for conditions on the phase composition and properties of Bi-HTSC revealed certain advantages of melt technology. Synthesis in the liquid phase presence allows to obtain more dense and low-porous target materials [1]. When melt quenching, it is possible to obtain a material in the amorphous state [2, 3], which determines the prospect of grain size adjustment, a higher rate of interfacial interactions due to the contribution of the metastable initial condition, and the material increased density. In addition, such a method is more likely than a solid phase method to assume the possibility of obtaining a textured material [4, 5].

The results comparative analysis of the use of different methods for the synthesis of oxide materials (including oxide superconductors) in the melt shows certain advantages of using a concentrated radiant flux as the heating source [6–9]. With respect to the fulfillment of the requirements determined by the specificity of the peritectic reaction for the formation of HTSC compositions 2223, 2234, 2245, a positive argument of this approach is the possibility of maintaining and rapidly adjusting the stationary mode of synthesis in the presence of the liquid phase. The positive contribution is: the purity of the process, carrying out the process in a crucibleless way, the completeness of the reactions and the best homogenization in the melt, carrying out the synthesis at an equilibrium oxygen partial pressure, the inertialess control of the amount of the liquid phase, the possibility of creating a texture due to a sharp temperature gradient at the melt-solid interface, directed crystallization at a given rate of cooling of the melt, the realization of the glassy state and control of the amount of the crystalline phase of the irovaniem cooling rate [8] .

The article purpose is to investigate the effect of melting, quenching and heat treatment conditions on the completeness of the formation of BSCCO nominal compositions 2223, 2234, 2245.

2. Experimental method

As a heating source, a concentrated radiant flux, created by a 10 kW xenon lamp, was used.

The initial charge of nominal compositions corresponding to $Bi_{1,7}Pb_{0,3}Sr_2Ca_{n-1}Cu_nO_y$ (n = 3, 4, 5) was prepared from oxides of Bi2O3, PbO, SrCO3, CaO, CuO. Beam samples melting was carried out in a crucibleless manner on a water-cooled substrate. For quenching, the melt dispersion technique was used.

The phase composition was monitored by X-ray diffraction using a DRON-UM1 diffractometer, CuK_α radiation and Diffractometer Model Rigaku Co, Ltd., Tokyo, Japan. The microstructure was examined with Neophot and TEM microscopes, Model JEM-1200EX11, JEOL Co., Ltd., Tokyo, Japan). The critical temperature was measured by a four-contact method.

3. The obtained results and discussion

As a result of quenching, samples of three types were obtained: pieces measuring about 0.5 cm3, plates 0.1 – 0.3 mm thick; Needles 1 - 10 mm in length and 0.1 – 0.4 mm in diameter. In the initial state after quenching, the phase composition of the pieces and plates was represented by amorphous and crystalline phases. The needles were completely amorphous.

The hardened precursors were heat treated at a temperature of 840–845 °C, 18–25 hours. After that, complete crystallization of the glass phase occurred in all the samples and the predominant formation of a phase of composition 2212 with unit cell parameters a = 0.5427 nm, c = 0.5412 nm, c = 0.3135 nm. From the described precursors, ceramic samples were prepared using traditional ceramic technology.

The samples microstructure of nominal composition 2223 (as an example) is shown in Figure 1*a,b*. The nature of crystallization of phases in samples-pieces and plates was similar. Their structure was characterized by the formation of blocks of about 120 – 140 µm in size, consisting of tightly packed plates (Figure 1*a*). Needles had a slightly different structure due to the formation of blocks of much smaller sizes, equal to 5 - 8 µm (Figure 1*b*).

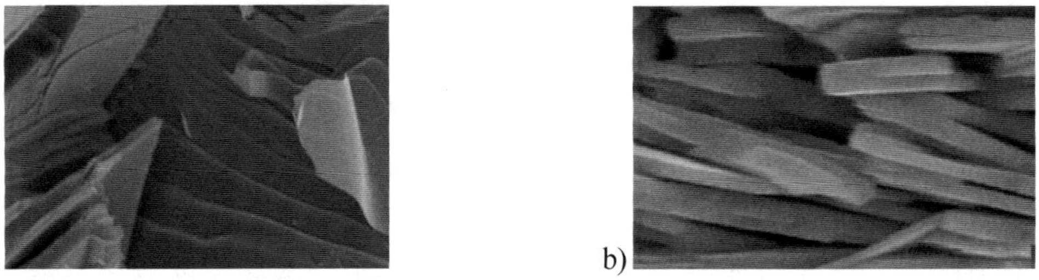

a) b)

Figure 1. Samples microstructure of the initial composition 2223; plates (*a*) and needles (*b*) obtained by ultrafast quenching of the melt and heat treated at 845 °C, 25 hours.

Synthesis of high-temperature superconducting phases was carried out by thermal treatment of ceramics samples in the form of a cylinder in the isothermal regime in the temperature range 845 – 850 °C in the interval 40–150 hours. According to X-ray analysis of samples [10], the main phase is 2223 (Figure 2). In addition to X-ray reflections related to phase 2223, additional reflexes are present in samples 2234 and 2245.

The parameters of elementary cells are determined and the features of the crystal lattice formation in dependence on the temperature-time parameters are investigated. The optimum time for formation of the superconducting phase in all the samples studied was 60–120 hours.

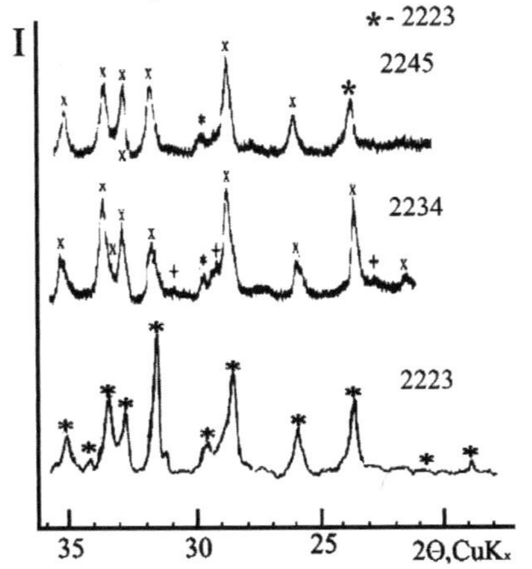

Figure 2. Samples diffractogram of a high-temperature superconductor of the nominal composition $Bi_{1,7}Pb_{0,3}Sr_2Ca_{n-1}Cu_nO_y$ (n = 3, 4, 5).

In the nominal composition samples corresponding to phases 2234 and 2245, a high-temperature modification is established, characterized by an identical crystalline structure, presumably tetragonal. These results allow one to assume a continuous mutual solubility of these phases at high temperatures [11].

The low-temperature form in the BSCCO samples of composition 2234 and 2245 corresponded to the ICDD data for 2234 with a monoclinic crystal structure.

All samples with a well-formed crystal structure were observed by the Meissner effect. The critical transition temperature for BSCCO samples of composition 2223 is determined to be 119 °K (Figure 3).

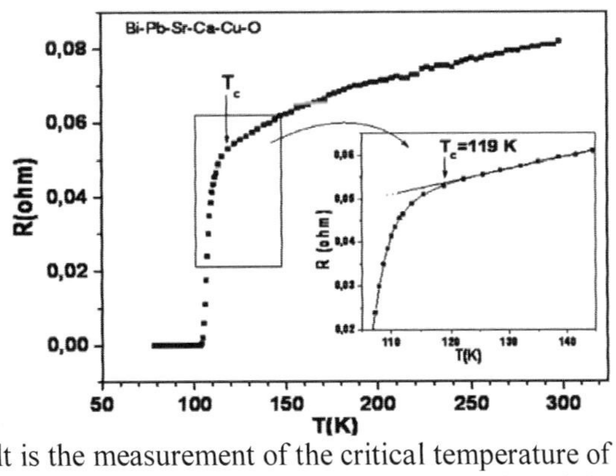

Figure 3. The result is the measurement of the critical temperature of the transition to the superconducting state of a sample of a superconducting ceramics of the nominal composition $Bi_{1,7}Pb_{0,3}Sr_2Ca_2Cu_3O_y$ in an electromagnetic field H = 250 Oe.

For the samples of superconducting ceramics of nominal compositions Bi1.7Pb0.3Sr2Can-1CunOy (n = 4, 5), the main phase was the superconducting phase of 2223, also had a transition to the superconducting condition of 107–110 K. However, traces of high-temperature superconducting phases up to 150 K were detected.

4. Conclusions

Based on amorphous precursors, a high-temperature superconducting ceramics of the nominal composition $Bi_{1,7}Pb_{0,3}Sr_2Ca_{n-1}Cu_nO_y$ (n = 3, 4, 5) was synthesized under the influence of a concentrated radiant flux. With an increase in the content of Ca and Cu, the temperature regime of the synthesis of superconducting phases decreases. It is established that for all synthesized samples the main phase is the superconducting phase $Bi_{1,7}Pb_{0,3}Sr_2Ca_2Cu_3O_y$, which has a critical temperature of the transition to the superconducting condition of 107 K. It is assumed that in samples of the composition $Bi_{1,7}Pb_{0,3}Sr_2Ca_{n-1}Cu_nO_y$ (n = 4, 5) is present phase with T_c exceeding 107 K.

References

[1] Skorykh V J *et al* 2001 5th Korea-Russia International Symposium on Science and Technology - Proceedings: KORUS **1** 42–45 doi: 10.1109/KORUS.2001.975049
[2] Goldshtein A E, Vavilova G V, Mazikov S V 2016 *Journal of Physics: Conference Series* **671** 012062 doi: 10.1088/1742-6596/671/1/012062
[3] Vavilova G V, Gol'dshtein A E 2018 *Measurement Techniques* **61**(3) 278–283 doi: 10.1007/s11018-018-1421-6
[4] Plotnikova I, Efremova O, Chicherina N, Dalinger V 2018 *MATEC Web of Conferences* **155** 01052 doi: 10.1051/matecconf/201815501052
[5] Gulamova D D, Uskenbaev D E, Fantozzi G, Chigvinadze J G, Magradze O V 2009 *Techn. Phys.* **6** 860–864
[6] Uskenbayev D E, Nogay A S, Aynakulov E B 2016 *IOP Conf. Ser.: Mater. Sci. Eng.* **110** 012030
[7] Gulamova D D, Uskenbaev D E 2006 *Applied Solar Energy* B **42** 40–43
[8] Gulamova D D, Uskenbaev D E, Turdiev Z S, Chigvinadze D G and Magradze O V 2009 *J. Applied Solar Energy B* **45** 196–200
[9] Plotnikova I V, Chaykovskaya O.N Sokolova I V, Artyushin V R 2017 *Journal of Physics: Conference Series* **881** 012006 doi: 10.1088/1742-6596/881/1/012006
[10] Plotnikova I V, Chicherina V V, Bays S S, Bildanov R G, O Stary 2018 *IOP Conference Series: Materials Science and Engineering* **289** 012029 doi:10.1088/1757-899X/289/1/012029
[11] Obukhov S G and *et al.* 2017 *IOP Conference Series: Materials Science and Engineering* **189** 012008 doi:10.1088/1757-899X/189/1/012008

IOP Conf. Series: Materials Science and Engineering **457** (2018) 012017 doi:10.1088/1757-899X/457/1/012017

The method of stabilization and reduction of noise in the measurement signal through the "dry" electrodes for electrocardiography

A Boyakhchyan[1], I Leznina[2], K Overchuk[1], A Uvarov[3]

[1]Postgraduate student, Tomsk Polytechnic University, School of Non-Destructive Testing, Tomsk, Russia
[2]Assistant professor, Tomsk Polytechnic University, School of Non-Destructive Testing, Tomsk, Russia
[3]PJSC MTS, Innovation Center, Moscow, Russia
E-mail: bojahchyan@yandex.ru

Abstract. The last 10 years have been actively developed for "dry" electrodes. Such electrodes should simplify the use of portable medicine. But scientists around the world are faced with difficulties in development. The article presents arguments about how to improve existing sensors.

1. Introduction

After the adoption of the law on telemedicine in the Russian Federation, which allows for remote medical diagnostics in 2018, telemedicine in the country began to develop, but still lags far behind foreign competitors.

The main goal is to achieve reliability of the data on human health using telemedicine gadgets (ECG, etc.). Doctors are skeptical of telemedicine gadgets without medical certification or clinical trials.

A group of young scientists of the Tomsk Polytechnic University were able to create a portable cardiograph and determine the diagnostic value of indications [1].

2 Materials and methods

Previously, the authors have already obtained the following results:

- tests of the prototype of the electrocardiograph in people with heart disease were carried out in order to determine the diagnostic value of the electrocardiogram obtained by recording the signal from the "dry" electrodes;
- a measuring system was manufactured and tested using the unique “dry” electrodes of the company Plessey Semiconductors;
- developed a special system for fixing electrodes, for better contact with the patient's body;
- the application field was identified - remote individual ECG monitoring systems (ST segment abnormalities, as well as cardiac rhythm disorders) in patients with coronary heart disease (CHD).

In the cardiograph were used “dry” electrodes from Plessey Semiconductors, which enable non-contact measurement without prior preparation (lubrication of the skin with contact gel) and additional sensors. Due to this, the measurement time is reduced, and the housing design provides ease of use and correctness of the imposition of electrodes.

The need to develop contactless electrodes appeared not so long ago (less than 10 years ago). So far, only Plessey Semiconductors is developing similar electrodes for sale. But since 2016, due to the sanctions imposed, the import of such electrodes into the Russian Federation has been banned, which at the moment makes their use in cardiographs impossible.

The assessment of the availability of diagnostic information in the signal obtained from the “dry” embedded electrodes was originally made on the basis of a comparison of the original electrocardiogram (ECG) obtained using the standard method, i.e. using chest leads using standard medical "wet" self-adhesive electrodes, with an ECG obtained using the built-in electrodes.

The study was conducted by the Research Institute of Cardiology of Tomsk. The research results are presented in the article [2]. Cardiologists of the Research Institute of Cardiology, as well as patient tests confirmed the diagnostic value of the ECG obtained from non-standard sensors. It was decided to create their own analogs of capacitive electrodes, while implementing the scheme in such a way as to avoid the disadvantages of the electrodes of the company Plessey Semiconductors.

3. Results and discussion

The main disadvantage of “dry” electrodes is their susceptibility to interference, which leads to the need to use additional means to improve the quality of the received signal, and this will also expand the possibilities of using the device in various conditions [3]. In order to determine which signal components need to be removed, was performed a spectral analysis of the signal. The results are presented in Figure 1.

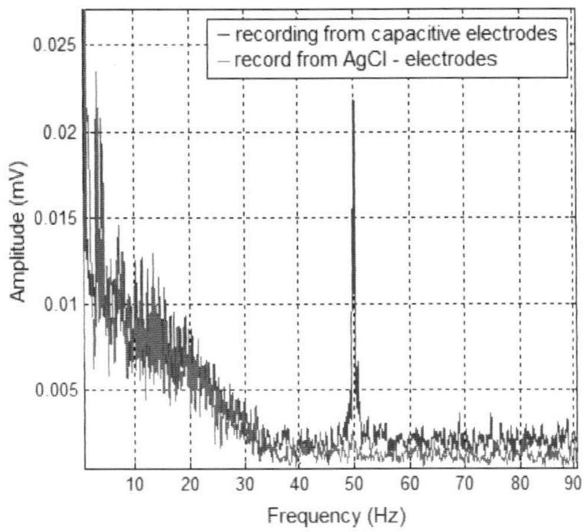

Figure 1. Spectral composition of signals.

Analyzing the obtained spectra, can conclude that during the use of capacitive electrodes two interferences appear in the signal:

- 50 Hz interference («power supply noise" or interference from power networks);
- broadband interference at a frequency of 10 Hz to 20 Hz.

50 Hz interference is removed from the signal in two ways:

- using a frequency filter;
- perform a complete shielding of the "dry" electrode and provide better alignment of signals between the human body and the device layout.

Broadband interference at frequencies of 10–20 Hz cannot be filtered by a frequency filter, since the spectrum of this interference overlaps the spectrum of the useful signal and the use of a frequency filter will lead to signal distortion, which will subsequently lead to incorrect interpretation and, as a result, incorrect diagnosis.

A review of the literature in the field of this problem was conducted, in consequence of which it was hypothesized that a solution could be found in recording the bioimpedance between the electrode and human skin. The studies were carried out with the help of the EKG-Express personal cardiograph (Figure 2 shows the "model" version of the device). Figure 3 shows the correlation dependence of bioimpedance and ECG.

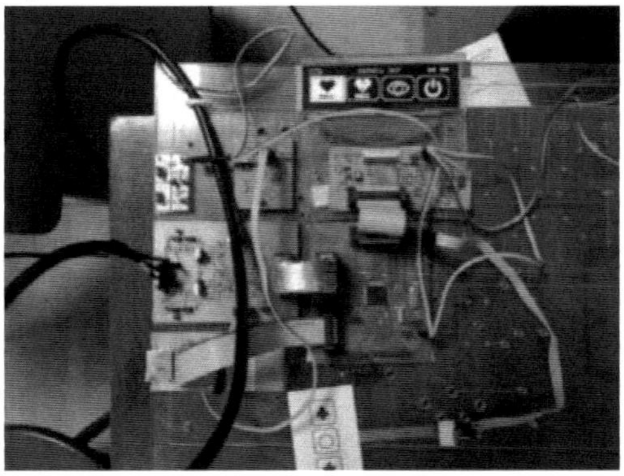

Figure 2. Layout for experiments.

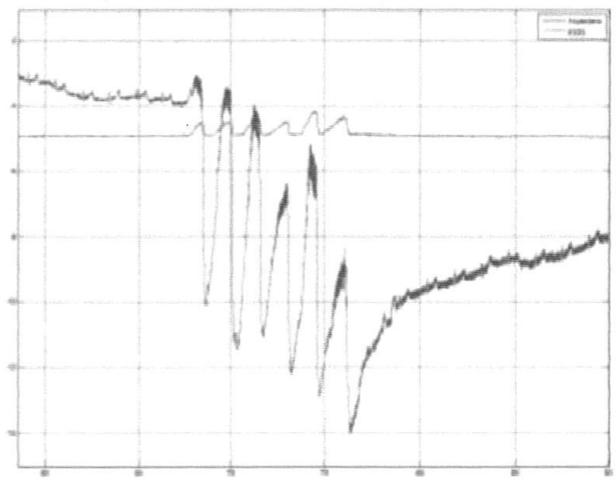

Figure 3. Registration of bioimpedance and ECG.

4 Conclusion

From Figure 3 it can be seen that the noise present in the ECG signal correlates with the change in bioimpedance. Based on this, it can be concluded that it is possible to create an adaptive system capable of isolating and subtracting interference from an ECG signal based on a signal from bioimpedance, thereby producing its stabilization. The next stage of work is the development of a new generation of capacitive electrodes for personal electrocardiography with detuning from the effect of the capacity of the skin-electrode contact.

Acknowledgments
The reported study was funded by RFBR according to the research project № 18-38-00535.

References

[1] Uvarov A A, Lezhnina I A, Overchuk K V, Starchak A S, Akhmedov Sh D 2016 *Journal of Physics: Conference Series* **671**(4) 012032 doi:10.1088/1742-6596/671/1/012032
[2] Boyakhchyan A, Lezhnina I, Overchuk K, Perchatkin V, Lvova A, Alexander U 2018 *IOP Conf. Series: Materials Science and Engineering* **363** doi:10.1088/1757-899X/363/1/012030
[3] Chi Yu M, Deiss S R, Cauwenberghs G *2009 Sixth International Workshop on Wearable & Implantable Body Sensor Networks Conference* 246–250 doi:10.1109/BSN.2009.52

IOP Conf. Series: Materials Science and Engineering **457** (2018) 012018 doi:10.1088/1757-899X/457/1/012018

Measurement of molecular humidity and oxygen in natural gas

I A Kondratyev[1], A A Shakhmatov[2], V P Pirog[3], L F Nosenko[4], E A Nikitina[5], P A Luchnikov[6], A A Nazarenko[6]

[1]Chief engineer, Experimental design bureau of automation, Angarsk, Russia
[2]Chief of department, Experimental design bureau of automation, Angarsk, Russia
[3]Associate professor, Experimental design bureau of automation, Angarsk, Russia
[4]Engineer, Experimental design bureau of automation, Angarsk, Russia
[5]Associate professor, MIREA – Russian Technological University, Moscow, Russia
[6]Researcher, MIREA - Russian Technological University, Moscow, Russia
E-mail: fisika@mail.ru

Abstract. The coulometric method of measuring humidity and oxygen on the basis of an electrolytic cell with a film of phosphorus oxide for continuous analysis of natural gas in pipelines of its transportation is considered. On the basis of the method the device is implemented that is designed to determine the mass concentration of water vapor $mg \cdot m^{-3}$, the dew point temperature in water at absolute pressure of the analyzed natural gas, as well as the volume fraction of oxygen. Measurements of the mass concentration of water vapor is from 1 to 100 $mg \cdot m^{-3}$, and the volume fraction of oxygen is from 10 to 10000 ppm at a gas pressure from 0.16 to 25 MPa. The main reduced measurement inaccuracy is from 3 to 6 %.

1. Introduction

Natural gas used as fuel and industrial raw materials in the production of various organic substances. Water vapor contained in the gas can condense under certain conditions and during its transportation in the pipeline (especially in winter conditions) leads to the formation of ice deposits, which leads to disruption of gas transportation systems. To solve the problem of humidity control in the gas, there are a number of methods along with which it is possible to distinguish the coulometric method.

2. Coulometry of gases

The coulometric method is based on the principle of absorption of moisture by the sorbent (a film of the substance of the sensing element) at a certain dosed flow of the analyzed gas, as well as electrolytic decomposition of this moisture into hydrogen and oxygen. The amount of moisture in the gas is determined by the electric current of the electrolysis process and the amount of gas flow.

The implementation of the coulometric method is carried out by means of a moisture electrolytic cell with a sensitive element (SE) from the sorbent film when a constant flow of the analyzed gas is passed through the coulometric electrolytic cell (CEC) [1, 2].

In SE cells GEC1 as a sorbent is used, the film partially hydrated phosphoric anhydride P_2O_5. A direct current (DC) voltage is applied to the cell electrodes. In SE cells are continuously taking place two processes: the almost complete absorption of moisture with a film of phosphoric anhydride P_2O_5

with the formation of phosphoric acid and electrolytic decomposition of water into hydrogen and oxygen with regeneration of phosphoric anhydride according to the reactions:

$$P_2O_5+H_2O \rightarrow 2HPO_3 \text{ and } 2HPO_3 \rightarrow H_2+\tfrac{1}{2}O_2+P_2O_5. \quad (1)$$

At a constant gas flow rate according to Faraday's law, the value of the electrolysis current is a measure of the moisture content of the gas. In this case, all the moisture in the flow of the analyzed gas is practically absorbed by the SE cell. The absorption of moisture and its electrolysis by the film of phosphorus oxide of SE cell occurs when an external DC voltage is applied to its electrodes, the value of which exceeds the decomposition potential of water. When operating in continuous operation, such a coulometric electrolytic cell performs electrolytic regeneration of phosphorus oxide SE, which provides a long service life of the cell.

3. The system of preliminary gas treatment

For pre-purification of natural gas from mechanical particles, droplet particles of oils, hydrogen sulfide, ethylene glycol and other organic compounds interacting with phosphorus anhydride SE in the measuring device implemented gas treatment system (GTS) (Figure 1).

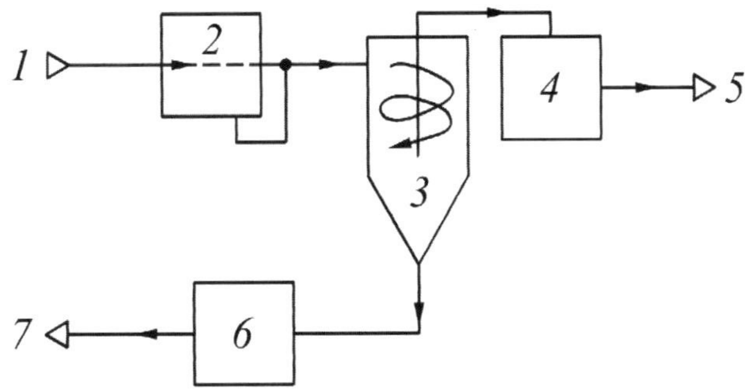

Figure 1. Block diagram of the primary gas flow preparation system GTS: 1 – gas input; 2 –pressure stabilizer; 3 – dynamic filter; 4 – fine filter; 5 – pure gas outlet; 6 – throttle; 7 – outlet to the collector drainage.

The work of the GTS (Figure 1) is to reduce the analyzed gas pressure stabilizer 2 and its purification from impurities. Block 2 provides an accurate output gas pressure within a predetermined value of ±30 %, adjusts the output pressure and reduces the influence of sorption processes in the volume of the gas path. The gas after the gas pressure stabilizer 2 enters the dynamic filter 3, where due to centrifugal forces it is cleaned from mechanical and droplet particles, and is divided into two streams. The purified gas through the filter 3 enters the fine filter 4, where heavy hydrocarbons and compressor oil vapor are removed. Further, the purified gas stream 5 is fed to the input of the analyzer (Figure 2). The second part of the impurity flow through the choke 6 is diverted to the drainage line 7 of the collector. This impurity part of the gas comes out at a flow rate of up to 2000 $cm^3 \cdot min^{-1}$ (at atmospheric pressure).

The lighter components of natural gas from the dynamic filter 3 are lifted up by a small flow through the filter 4 with a nominal flow rate of 50 $cm^3 \cdot min^{-1}$, which then enters the gas analyzer input into the first CEC1.

In blocks 2 and 3 and 4 of the primary gas flow preparation system (Figure 1) used not sorption thin organic coatings based on fluoropolymers and Parylene F, are applied to the inner surface elements and nodes from an active gas phase in a low vacuum [3, 4]. The coating thickness is 100÷1000 nm. The wettability angle of the surface on the water of such coatings is not less than 95 degrees.

4. Gas analyzer of moisture and oxygen

The gas analyzer implemented on the basis of a coulometric electrolytic cell includes a gas treatment system GTS unit and the analyzer of molecular water and oxygen content (Figure 2). Here, natural gas is supplied to the inlet A of the GTS (Figure 1) block where it is divided into two streams: the separation of moisture from impurities and the gas flow for molecular analysis at the inlet C.

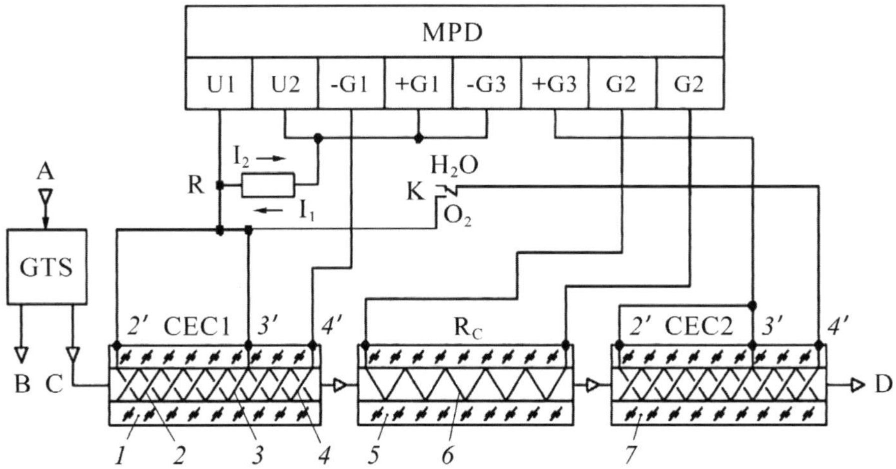

Figure 2. Block diagram of the gas analyzer: GTS – gas treatment system; MPD – microprocessor device; U1 and U2 – input information of ΔI in the microprocessor; G1 and G3 – direct current sources; G2 – alternating current source; R – precision resistor; K – measuring mode switch; R_C – reactor; CEC1 and CEC2 – coulometric electrolytic cell; 1, 5 and 7 – glass tube; 2', 3' and 4' – working, control and common electrodes, respectively; 6 – platinum coil heater.

Next, the analyzed natural gas with a constant flow rate passes through the first coulometric electrolytic cell CEC1 with an internal diameter of the channel 0.8 mm, and then through the R_C reactor enters the second same coulometric electrolytic cell CEC2. In the channel of the cylindrical glass tube 1 cell A1 electrodes are placed, made in the form of non-contacting spirals of radium: working 2', control 3' and total 4' contact electrodes. On the inner surface of the tube channel 1 is an electrode 3 with a film of sorbent-partially hydrated phosphorus oxide P_2O_5, which has a high sorption capacity of water molecules.

Under normal operating conditions, the moisture in the natural gas is absorbed by the SE sorbent film in cell CEC1 to the electrodes of which a DC source G1 is connected. The applied constant voltage exceeds the water decomposition potential. Simultaneously with the absorption of moisture in the film, a continuous electrolysis occurs with the formation of oxygen and hydrogen according to the reactions (1). In the steady-state mode, the amount of absorbed and decomposed moisture per unit time is equal and, consequently, the electrolysis current I_1 of the coulometric electrolytic cell CEC1 is a measure of the moisture content in the natural gas and is determined:

$$I_1 = B_{H2O} \cdot Q \cdot P_\partial / K_1 \cdot T, \qquad (2)$$

where I_1 – current electrolysis of water in GEC1, mkA; B_{O2} – volume fraction of moisture in the analyzed gas, ppm; Q – flow rate of the analyzed gas, $cm^3 \cdot min^{-1}$; P_∂ – barometric pressure, kPa; T – ambient temperature, K; K_1 – normalizing coefficient due to the choice of units of physical quantities in the measurement of moisture, ppm $cm^3 \cdot kPa \cdot K^{-1} \cdot min^{-1} \cdot mkA^{-1}$.

The current I_1 flows in this direction through the precision resistor R, which creates a voltage drop recorded by the microprocessor device MPD. From the outlet of the first cell CEC1, the hydrogen, oxygen and dry natural gas formed enter the reactor R_C. In the cylindrical channel of the glass tube 5 of the reactor R_C, a platinum spiral of the heater 6 is placed, connected to the alternating current source G2. Under the action of the heated catalyst 6, the reaction of the interaction of the micro-impurities of oxygen with hydrogen with the formation of water molecules takes place.

Then natural gas is obtained in the catalyst micro preemies moisture enters the second similar to the first cell of the CEC2 in a glass tube 7 with the inner channel diameter of 4.0 mm. Electrodes of the cell are made of platinum-iridium wire and connected to the constant current source G3. Through the output D, the output of the analyzed gas at atmospheric pressure is provided.

5. Experimental result

In the steady-state operation of the gas analyzer (Figure 2) the amount of moisture absorbed and decomposed per unit time is equal. In this case, the electrolysis current of the coulometric electrolytic cell is a measure of the total volume fraction of moisture in the natural gas to the CEC1 cell, and moisture additionally formed in the R_C reactor due to the interaction of natural gas oxygen to GEC1 and the resulting trace oxygen with hydrogen.

The electrolysis currents of I_1 cell CEC1 and the electrolysis current of I_2 cell CEC2 in the mode of measuring the volume fraction of oxygen O_2 flow through the precision resistor R in opposite directions, creating a difference current $\Delta I = I_2 – I_1$. The value of the current ΔI is due to the current of electrolysis of moisture, which was formed in the reactor R_C in interaction with the oxygen of the analyzed natural gas. The difference current ΔI is a measure of the oxygen concentration in natural gas in the mode of measuring the volume fraction of oxygen O_2 and is determined by the formula:

$$\Delta I = I_2 – I_1 = B_{O2} \cdot Q \cdot P / K_2 \cdot T, \quad (3)$$

where I_1 – current electrolysis of water in GEC1, mkA; B_{O2} – volume fraction of moisture in the analyzed gas, ppm; Q – flow rate of the analyzed gas, $cm^3 \cdot min^{-1}$; P – barometric pressure, kPa; T – ambient temperature, K; K_1 – normalizing coefficient due to the choice of units of physical quantities in the measurement of moisture, ppm $cm^3 \cdot kPa \cdot K^{-1} \cdot min^{-1} \cdot mkA^{-1}$.

Table 1 shows the results of experimental studies of the molecular moisture content in natural gas using the element-by-element method of verification.

Table 1. Experimental data of volume fraction of moisture in natural gas (VFM)

№	Absolute gas pressure, MPa	I_1, мкА	VFM, ppm	VFM, mg/m^3	$^oS_{t.r}$
1	2.2	4483	670.6	503	11.9
2	2.6	2505	374.7	281	5.8
3	3.0	137.7	20.6	15.5	-25.0
4	3.4	474.6	71.0	53.3	-11.3
5	3.8	4171	623.9	468	10.7
6	4.2	1473	220.3	165	4.3
7	4.6	35.7	5.3	3.98	-32.7
8	5.0	846.4	126.6	95.0	-0.6
9	5.4	259.9	38.8	29.1	-13.0
10	5.8	71.24	10.7	8.03	-24.7
11	6.2	19.99	3.0	2.25	-34.8

The volume fraction of moisture and oxygen is determined after appropriate transformations by the microprocessor device and displayed on a digital display indicating the type of gas (H_2O or O_2).

The performance of the system sample gas is provided at pressures up to 25 MPa. The dew point temperature of the moisture $^oS_{t.r}$ is from minus 69.4 to plus 31.3 °C with the basic absolute error ±2 °C, and in the measurement range from minus 69.4 °C to minus 20 °C is ±1.5 °C.

Performed analyzer in ppm at atmospheric ambient pressure values $^{o}S_{t.r}$ when the pressure in pipeline transport. The deviation of the real gas humidity from the humidity of its ideal model increases with decreasing temperature and increasing pressure [5].

Figure 3 shows graphs of experimental calibration data of the gas analyzer when measuring the concentration of oxygen N_{O2} (curve 1) in natural gas, using the element-by-element method of verification and the corresponding difference current ΔI (curve 2) in the precision resistor R of the measuring circuit (Figure 2).

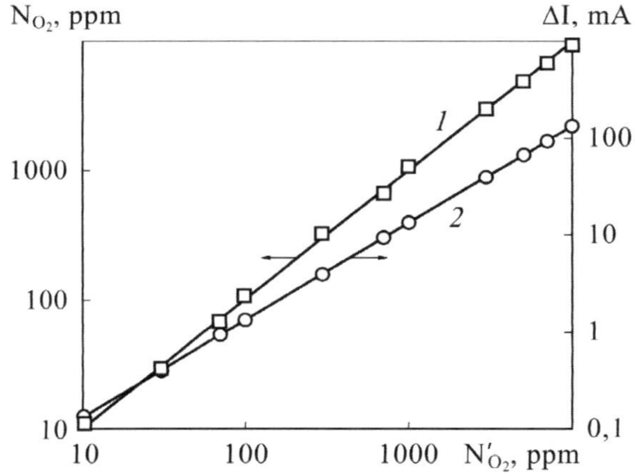

Figure 3. Graphs of gas analyzer calibration data for measurement of oxygen N_{O2} (1) concentration in natural gas and difference current ΔI (2) in the precision resistor.

6. Conclusion

The method of coulometric electrolytic cell is implemented when creating a gas analyzer of moisture and oxygen of natural gas, which is designed to measure: the mass concentration of water vapor mg/m^3, the dew point temperature of the water at the absolute pressure of the analyzed gas, as well as the volume fraction of oxygen. The gas analyzer in the hygrometer mode measures the mass concentration of water vapor from 1 to 100 mg·m^{-3}, the volume fraction of oxygen from 10 to 10000 ppm at a gas pressure from 0.16 to 25 MPa. The main reduced error (to the upper value of the measurement range) from ±4 to ±6 %. The dew point temperature of the moisture $^{o}S_{t.r}$ is from minus 69.4 to plus 31.3 °C with the basic absolute error ±2 °C, and in the measurement range from minus 69.4 °C to minus 20 °C is ±1.5 °C.

Acknowledgments

This work was done at MIREA – Russian Technological University and was supported by the Ministry of Education and Science of the Russian Federation (project № 3.8074.2017/8.9 basic part of the State task).

References

[1] Kondratyev I A *et al* 2018 *IOP Conf. Ser.: Mater. Sci. Eng.* **289** 012027 doi:10.1088/1757-899X/289/1/012027
[2] Klopotov K I *et al* 2010 *Coulometric electrolytic cell* Patent RF 2488107 Int. Cl. G01N 27/28 (in Russia)
[3] Luchnikov P A *et al* 2017 *IOP Conf. Ser.: Mater. Sci. Eng.* **189** 012014 doi:10.1088/1757-899X/189/1/012014
[4] Luchnikov P A 2018 *IOP Conf. Ser.: Mater. Sci. Eng.* **289** 012037 doi:10.1088/1757-899X/289/1/012037
[5] Pirog V P *et al* 2011 *Meas. Tech.* **53** 1411 doi:10.1007/s11018-011-9674-3

Current converters with improved power characteristics

T B Akimzhanov[1], S K Zhumazhanov[2], Ye V Ivanova[3], B S Masabaev[4]

[1]Senior lecturer, Department of Operating Electric Equipment, Saken Seifullin Kazakh Agrotechnical University, Astana, Kazakhstan
[2]Senior lecturer, Department of Operating Electric Equipment, Saken Seifullin Kazakh Agrotechnical University, Astana, Kazakhstan
[3]Professor, Department of Power Engineering, Metallurgy and Information Technologies, Innovative University of Eurasia, Pavlodar, Kazakhstan
[4]Master, Department of Operating Electric Equipment, Saken Seifullin Kazakh Agrotechnical University, Astana, Kazakhstan
E-mail: takimzhanov@mail.ru

Abstract. This article discusses development issue of semiconductor converters with the natural current external characteristics. The converter consists from a parametric current source, to the outputs of which are connected gated sets. The parametric source is a resonant circuit, which transforming electromotive force input system to the output system of currents. Here the resonant circuit in the current source provides an improvement in the power characteristics of the converter, so increasing energy efficiency indicators. This current converters use for supply most energy-intensive consumers with a steeply falling external characteristic (arc furnaces, electrolysis plants of non-ferrous metals, etc.), and also in electric drives with erection torque motor.

1. Introduction

Modern industrial production places ever increasing demands on sources and converters of electrical energy by reliability and cost. Electric energy is produced and distributed by alternating current at frequency of 50 Hz. At the same time more than 30 % of electricity is consumed at direct current [1]. The direct current is obtained by the conversion of alternating current by semiconductor converters. The power source for energy-intensive consumers (electrothermy, electrolysis plants of non-ferrous metals) is widely used by semiconductor converters with characteristics of current sources. For example, the current source for electrothermal plants allows the stabilization of the operating current and other indicators. Current sources in arc steel furnaces provide a reduction in metal consumption for smelting, eliminate the influence of the “short grid”, increase the power consumption (cosφ, efficiency, harmonic ratio, etc.) [2]. In electrolysis plants is increased the productivity of the technological process, which is directly proportional to the current. Existing semiconductor converters with current source characteristics have advantages and also disadvantages. They are caused by nonlinear elements (diodes, thyristors) in the circuit, which worsens the quality of electricity and energy efficiency of electrical networks [3–5]. Therefore, the development and creation of efficient current converters is an important and urgent task [6].

2. Analysis of the work of current transformers for different types of loads

2.1. Active load (contact heating)
Contact heating is one of the methods of direct heating. Direct heating in comparison with indirect has the following advantages, it allows to increase labor productivity, significantly reduce metal losses to scale, extend the life of equipment, improve working conditions, etc.

Contact heating provides a high heating rate (tens of seconds). This is achieved through intensive electrical connection. Industrial contact heating installations are characterized by high currents (several tens of kilo amperes) and small applied voltages (5…25 V). One of the main disadvantages of industrial contact heating installations is their relatively low energy performance (efficiency is not more than 75%, power is about 0.8). Besides, without additional equipment, the output current can not be smoothly controlled [2, 7, 8].

2.2. Active - inductive load (arc furnace)
Arc furnaces are powerful consumers. They generally operate with a low power factor (0.7…0.8). The electric mode of arc furnaces is characterized by frequent technological short circuits (5…10 per minute). Some arc furnaces (vacuum arc furnaces) operate on direct current. In this case, the power supply must include an alternating current to direct current converter.

Vacuum arc furnaces are very sensitive to regime changes. Current fluctuations do not affect the quality of the ingot, if these fluctuations do not exceed 3 - 5%. Therefore, it is necessary to stabilize the current of the arc furnaces and its program change (regulation) for the time of welding and heating (the beginning of the melting), and for the time of the withdrawal of the shrinkage shell (melting end) [2, 9, 10].

2.3. Load in the form of counter- electromotive force (electrolysis of non-ferrous metals)
The work of electrolyzers is described by current-voltage characteristics (CVC) with small slopes in the working zone [10, 11]. A small slope of the current-voltage characteristic in the working zone makes it necessary to automatically stabilize the load current, because even slight variations in the grids voltage lead to a significant change in the current. So, when the grids voltage is reduced by 10% of the nominal value, the current of the zinc electrolyzers can be reduced by 10 to 50% [1].

3. Single-phase current parametric converters (CPC) for various types of loads
One of the ways to solve the problem of increasing the power indexes of gate converters is the development of CPC. They are compared with the traditional types of semiconductor converters have a power factor close to one and the natural current external characteristics [11–14].

3.1. Single-phase CPC for active load
Single-phase schemes of current parametric converters can be assembled by bridge or zero schemes. They are used usually at a relatively low power. The shape of the current in the load in single-phase current parametric converters does not depend on the scheme of the gate set, but is determined by the nature of the load. The scheme and time diagrams of the operation of a bridge single-phase current parametric converter are shown in Figure 1 and Figure 2, where I – is a parametric current source. Gates 1 to 4 operate in the order of the natural unlocking and provide in the load rectification mode. Latitudinal regulation of the load current is carried out by changing the duration of the short-circuited mode ($I_H = 0$).

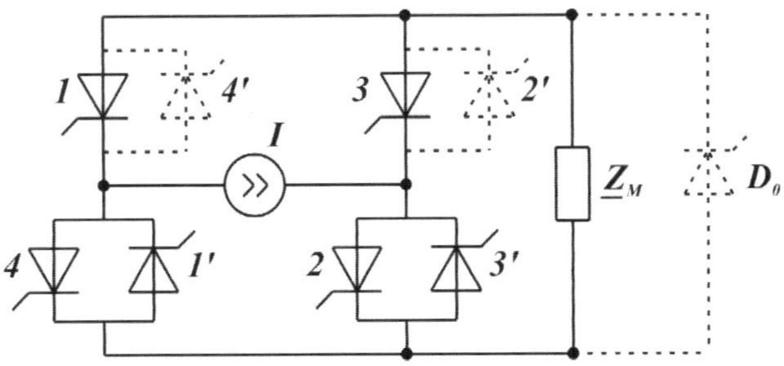

Figure 1. Single-phase scheme of CPC.

To ensure a short-circuited mode, two additional gates 1' and 3' are entered into the circuit. They are connected in parallel to gates 4 and 2. At the next clock cycle, instead of the gates 3 and 4 are included gates 3' or 4', the output of the parametric current source is short-circuited and the load current becomes zero. If it is necessary to adjust the duration of the pause, then at the next clock cycle includes gates 2, 1', and so on.

When the valves are turned on in the order of the natural unlocking, in the load restored a positive current value. Due to the characteristic features of the current parametric converters (the inadmissibility of idling, etc.), control impulses for the gates of CPC must be supplied only at the moment when the output current passes through zero.

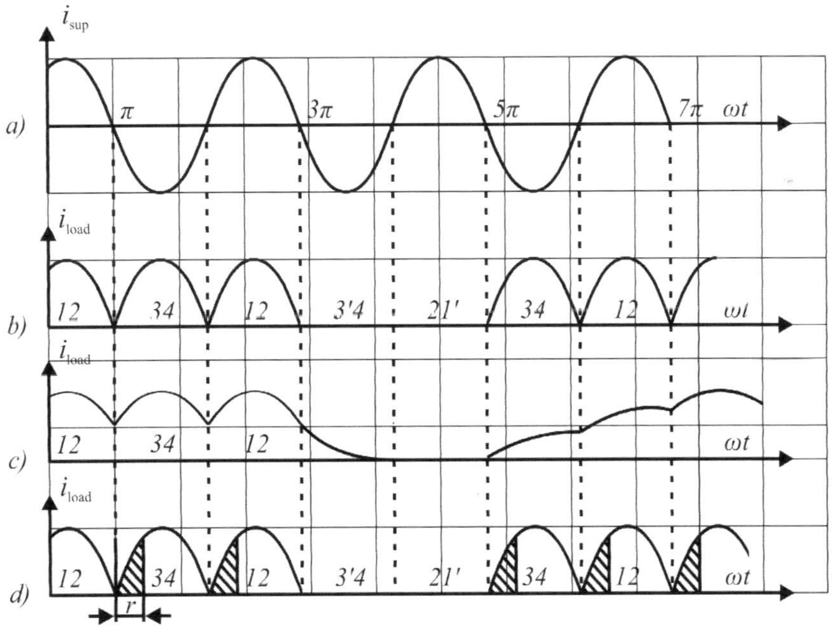

Figure 2. Time diagrams of single-phase current parametric converter operation (a) on the active load (*b*), on the active-inductive load (*c*), on the load in the form of counter-electromotive force (*d*).

Some consumers need a reverse load current [5]. To ensure the reversal of load current in the converter scheme are input two additional gates 2' and 4'. They are connected in parallel to the gates 3 and 1 (Figure 1). Then, at the next cycle, after a short-circuited mode (gates 2, 1' are turned on), the control pulses are applied to the converter gates 3', 4' and there is a smooth reverse of the load current.

3.2. Single-phase CPC for active-inductive load

The algorithms of operation of single-phase CPC on the active-inductive load are similar to those previously considered, when working on an active load. However, it is necessary to switch on an additional diode D_0 to provide a load-loop circuit in the short-circuited mode (Figure 1). The time diagrams of the load current of a single-phase CPC with an active-inductive load are shown in Figure 2*c*.

3.3. Single-phase CPC for loads in the form of counter-emf

A characteristic feature of the converter operation on the counter-emf is the presence of a commutation interval $\boldsymbol{\gamma}$ (Figure 2*d*), the duration of which essentially depends on the load value and is determined by the recharge time of the capacitance of the parametric source within the limits from $-E_d$ up to $+E_d$. In this case, the current in the load does not flow in the switching interval.

It should be noted, the mode $I_H = 0$ allows only one-sided closing of the load during it operates on a counter-emf, since the two-sided closure is emergency. Reversing the load current when the CPC is running on the counter-emf (that is equal to E_d and the constant E_d polarity) is inadmissible, since a loop for two-sided closing of the load is formed through the operating rectifier bridge getes. Therefore, the reversal of the load current is only permissible if the polarity E_d is reversed. Time diagrams of operation of a single-phase converter in the operation of a counter-emf are shown in Figure 2*d*.

4. Conclusion

The most energy-consuming consumers of direct current are the electric technological plants (arc furnaces, etc.), which have a non-linear current-voltage characteristic with a low differential resistance. The quality of control of power that transmitted from source to consumer depends on the coordination of their characteristics. In the optimal case, these characteristics must be opposite, therefore, for consumers with a low differential resistance, current converters are necessary.

These requirements correspond to the current converters described in the article. These converters are based on a parametric current source and connected to the output of the gated sets (current-to-parameter converters) with the possibility of latitudinal control of the output current of the load, which is confirmed by the time diagrams of their operation.

In arc steel smelting furnaces, that are characterized by high operating currents, the use of current sources provides power stability in process short circuits, which significantly reduces the metal loss for melting, eliminates the influence of the "short grid", improves the power quality indicators.

In vacuum arc furnaces, when using of the proposed current parametric converters, the quality of remelted ingots significantly improves, the explosion safety of furnaces increases.

In electrolysis plants and electrochemical machining plants, the application of the CPC makes it possible to increase the productivity of the technological process, since it is directly proportional to the current.

In addition to electrothermal plants, the use of sources of stabilized direct current for an electric drive (electric drive with erection torque motor) is perspective [13, 15–18]. The converters considered in this article can be used in the practical design of secondary power supplies for powerful consumers.

References

[1] Nikulin A D, Rodshtain L S and Salnikov V G 1995 *Thyristor Conversion Technique in Non-Ferrous Metallurgy* (Metallurgy Publ: Moscow)

[2] Illahi F, El-Amin I and Mukhtiar M U 2018 *TIEEE Transact. on power delivery* **33** 4 1727–1734

[3] Mendybayev S, Sarsikeyev Ye and Kapanova D 2018 MATEC Web of Conferences **155** 01047 doi: 10.1051/matecconf/201815501047

[4] Zhumazhanov S, Utegulov B, Utegulov A, Begentayev M, Uakhitova A and Zhakipov N 2011 *Proc. The IASTED Int. Conf. on "Modelling, Simulation, and Identification (MSI 2011)"* 147–150 doi: 10.2316/P.2011.755-030

[5] Akimzhanov T B, Kharlov N N, Borovikov V S and Ushakov V Ya 2014 *9th International Forum on Strategic Technology, IFOST 2014* 351–354 doi: 10.1109/IFOST.2014.6991138

[6] Hocine L, Yacine D, Kamel B and Samira K M 2009 *Energy* **34**(9) 1207–1214 doi: 10.1016/j.energy.2009.03.003

[7] Maki S, Harada Y, Mori K, Makino H 2002 *J Mater. Proces. Technol.* **125–126** 477–482 doi: 10.1016/S0924-0136(02)00324-2

[8] Ozturk F, Ece R E, Polat N, Koksal A 2011 *Key Eng Mater* **479** 130–136

[9] Lucia O, Maussion P, Dede E J and Burdio J M 2014 *IEEE Transact. on indust. electron.* **61**(5) 2509–2520 doi: 10.1109/TIE.2013.2281162

[10] Alonso M A P and Donsion M P 2004 *IEEE Transact. on power delivery* **19**(1) 367–373 doi: 10.1109/TPWRD.2003.820189

[11] Mendybayev S A and Ivanova Ye V 2012 *Proc. of Int. Conf. Efficient and qualitative use of electricity (in Russian)*

[12] Varadan S, Makram E B and Girgis A A 1996 *IEEE transact. on power delivery* **11**(3) 1685-1690 doi: 10.1109/61.517535

[13] Mendybayev S A, Isenov S S and Aibolatova A K 2016 *Proc. of Int. Conf. Technical and Mathematical Sciences (in Russian)*

[14] Obukhov S G, Sarsikeyev Ye Zh, Sholokhova I I 2016 *Proc. of 2nd International Conference on Industrial Engineering, Applications and Manufacturing, (ICIEAM 2016*) 7911454 doi:10.1109/ICIEAM.2016.7911454

[15] Ibraev K A, Sarsembieva E K 2017 *Proc. of Int. Siberian Conf. on Control and Communications (SIBCON 2017)* 7998492 doi: 10.1109/SIBCON.2017.7998492

[16] Mendybayev S A, Issenov S S, Issenov Zh S, Nurzhan N N 2017 *Proc. of Int. Siberian Conf. on Control and Communications (SIBCON 2017)* 7998502 doi: 10.1109/SIBCON.2017.7998502

[17] Plotnikova I, Vedyashkin M, Mustafina R, Plotnikov I, Tchaikovskaya O, Josefa Bastida, Verpeta M 2016 MATEC Web of Conferences **79** 01019 doi:10.1051/matecconf/20167901019

[18] Bemš J, Králík T, Kubančák J, Vašíček J, Starý O 2014 *Radiation Physics and Chemistry* **104** 398-403 doi:10.1016/j.radphyschem.2014.02.008

IOP Conf. Series: Materials Science and Engineering **457** (2018) 012020 doi:10.1088/1757-899X/457/1/012020

Calculation of macrodefects coordinates in dielectric specimens on the two-dimensional mathematical model of mechanoeletric transformations method

R A Laas[1], P N Khorsov[2], A P Surzhikov[3]

[1]Assistant, National Research Tomsk Polytechnic University, Tomsk, Russia
[2]Senior researcher, National Research Tomsk Polytechnic University, Tomsk, Russia
[3]Chief Researcher, National Research Tomsk Polytechnic University, Tomsk, Russia
E-mail:romanlaas@tpu.ru

Abstract. Two-dimensional mathematical model of dielectric specimen was used to determine the capability of mechanoelectrical transformations (MET) method to localize macrodefects. Amplitude and phase characteristics of response signal analytical representation were used as response parameters. Three different types of short radiofrequency pulses were chosen for the excitation. A short sin curve of a single period interval is most useful to search the position of the defect, whilst pulses of higher frequencies and are better for location depth evaluation.

1. Introduction

MET method is widely used to control structure imperfections of dielectrical materials [1,2]. It allows to determine value of stress-strain state [3,4]. In MET method acoustic excitation wave formed by impact device interacts with MET sources, such as double electric layers on the interfaces of different materials or enclosures with piezoelectric properties [5,6].

Excitation wave changes electric dipole moment of the source, which leads to the generation of alternating electromagnetic field. The parameters of the field can be registered by a transducer, so that signal reception does not require surface contact as distinct from ultrasonic methods. The modification of this method, which allows to significantly increase signal to noise ratio using multiple excitations of specimen without changing the source-specimen-detector system, substantively increase the efficiency and capabilities of the method [7].

Mathematical modelling using 1-dimensional models has shown the capability of the method to determine the depths of defects by phase analysis of analytic signal interpretation. Series of experiments has proved the consistency of the model [8].

Narrowband signal was obtained using high frequency pulse (100 kHz) with 1 ms duration as an excitation signal. The response time were chosen to be greater than time required for beam of acoustic excitation to reach the studied defect, reflect to the detector, so that it was possible to register the phase jump of interference between reflected wave and structural electromagnetic noise signal.

The goal of this paper is to evaluate the efficiency of surface scan with MET-method of localizing macrodefects and estimating their depth using wideband signals. Consider scanning system is a solid single module where piezoelectrical transmitter and capacitance transducer are installed, so that both emission and detection of signal take place on one side of the object.

2. The mathematic model

The control system is shown on Figure 1.

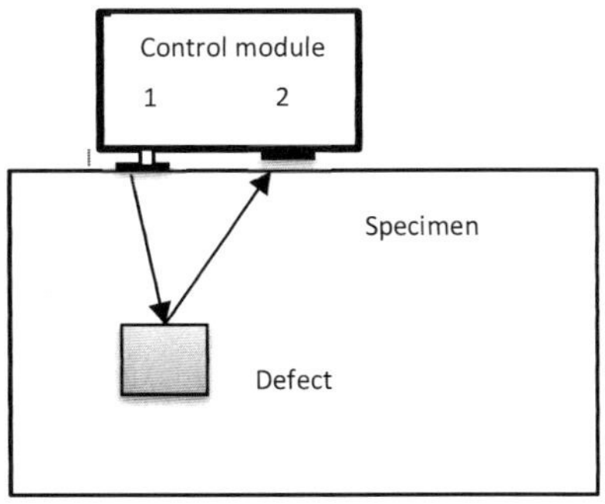

Figure 1. The control system.

On the Figure 1 numbers 1 and 2 mark surfaces of transmitter and transducer correspondingly. Arrows show acoustic excitation beam, reflected from the surface of defect toward transducer.

To describe 2-dimensional parameters of control module (CM) mathematical model was developed. It allows considering linear dimensions of transmitter and transducer, distance between them, allows determining frequency, shape and length of input signal and parameters of the defect, such as depth and size. The model allows adding a white noise of certain intensity to informative signal.

The algorithm of response parameters calculation is described below.

Beams of acoustic excitation wave are being reflected from the surface of defect toward transducer. Each of beams emitted from every discrete point of transmitter reaches every discrete point of defect surface. Levels of signals are calculated at this stage with the consideration of the translation distance for each beam. Then levels of each reflected beam signal are registered at each point of transducer. Step width can be found as follows:

$$\Delta s = \frac{v}{f_d}$$

where v – speed of sound in specimen; f_d – sampling rate.

Considering speed of sound 3000 m/s and sampling rate 1 MHz Δs = 3mm.

The distances $r1_{ijs}$ from each point of transmitter to each point of defect, facing control module, are calculated as follows:

$$r1_{ijs} = \sqrt{h^2 + [(x0+i) - (x1_s + j)]^2},$$

where h – depth of defect location; x0 – initial horizontal position of the defect; $x1_s$ – initial position of transmitter plate; s- control module shift index along the surface; i – beam shift index along defect surface (i=0, 1, …$b/\Delta s$), b – linear size of the defect; j – beam shift index along the surface of the transmitter; (j=0, 1, …$c/\Delta s$), c – linear size of transmitter plate.

$r2_{iks}$ is calculated analogously:

$$r2_{iks} = \sqrt{h^2 + [(x0+i) - (x2_s + k)]^2},$$

where $x2_s$ – initial coordinate of transducer plate; k – ray shift index along the surface of the transducer; (k=0, 1, …$m/\Delta s$); m – linear size of transducer plate.

The next step is to calculate levels of signals $U1_{ijs}$ in each point of the defect surface taking into account the angular pattern of the transmitter, which is chosen proportional to cos(φ1) where φ1 – angle between beam direction and defect surface normal. That is an appropriate approximation for relatively small angles and wavelengths comparing to size of the transmitter. The spherical nature of the wave has to be considered as well, this leads to the fact that level of signal A is in reverse ratio to the distance from the transmitter. As a result, the formula looks as follows

$$U1_{ijs} = A \cdot \frac{\cos(\varphi 1)}{r1_{ijs}} = A \cdot \frac{h}{r1_{ijs}^{\;2}}.$$

The total level of acoustic signal at the transducer can be found as:

$$U_{jks} = A \cdot \frac{h^2}{r1_{ijs}^{\;2} \cdot r2_{iks}^{\;2}}.$$

Time delay of the corresponding beam can be found as:

$$t_{jks} = \frac{r1_{ijs} + r2_{iks}}{v}$$

The corresponding indexes tt are to be found as:

$$tt_{jks} = round(t_{jks} \cdot f_d).$$

where round is an operator which rounds up to the nearest whole number. The sum of all signal levels with same time indexes is a pulse response curve U_{tt}. Let us set RF impulse G_{tt} of a given length with certain frequency. The result is a convolution R_{tt} of pulse response curves U_{tt} and G_{tt}.

$$R_{tt} = \sum_{\tau=0}^{tt} U_{tt-\tau} \cdot G'_{\tau},$$

where top index over *G* means derivative with respect to time.

3. Calculations

Assume certain parameters of the system: linear size of the defect is 20 mm, depth of location – 20 mm , linear size of transducer is 20 mm as well, distance between transmitter and transducer 5 mm. Let us compare efficiency of localization for different signals in this given geometry.

Three different types of signals were used: square-topped pulse with a length of 4.5 μs and radiofrequency impulses of the same length with frequencies 100 kHz and 200 kHz. Under such conditions one or two full periods are in a signal time window correspondingly.

Responses of given radiofrequency impulses were transformed into analytical signal using Hilbert-transformation as follows:

$$S(t) = f(t) + i \cdot h(f(t)),$$

where *h*() – Hilbert transformation operand, *i*- imaginary unit.

Then phase dependency on time φ(t) were calculated. It`s complicated to use this characteristic directly due to the rapid rise of phase linear time.

To obtain residual phase change from time the subtraction procedure was performed for linear phase component in the specified time interval by the formula:

$$\varphi_1(t) = \varphi(t) - \varphi(t_0) - \frac{(\varphi(t_1) - \varphi(t_0)) \cdot (t - t_0)}{t_1 - t_0},$$

where $\varphi_1(t)$ is the current value of phase on time dependency, φ(t0) is the value at the initial time t0 of the given interval, $\varphi(t_1)$ is the value at the final time t_1.

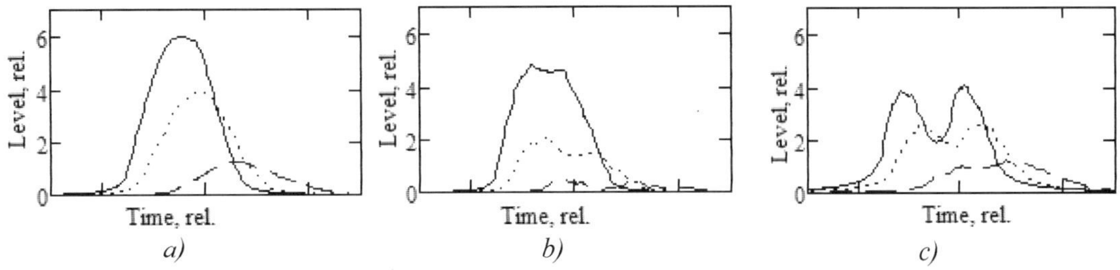

Figure 2. Amplitude-temporal characteristics of analytical representations: *a*) RF 100 kHz, *b*) – 200 kHz, c*)* square-top.

Figure 2 shows amplitude-temporal characteristics of response signals analytical representations after the impact of signals with chosen shapes under different positions of Control Module. Figures 2*a*,*b*,*c* are responses for 100 kHz, 200 kHz and square-top impulses correspondingly. The continuous line corresponds to the position of CM right on top of the defect; dashed line represents the shift by 10 mm; stripped line – by 20 mm.

It is seen that RF signal of 100 kHz gives response of highest level, square-topped – lowest level. The time-shift of impulses with the distance can be explained by growth of travelling lengths for each excitation beam. Offset-dependent amplitude attenuation is due to scattering of sphere-wave in material. This model ignores inner friction related signal attenuation as well as rounding the defect that leads to reduction of reflected toward transducer signal level.

Further development of the model is planned to be considering these current limitations.

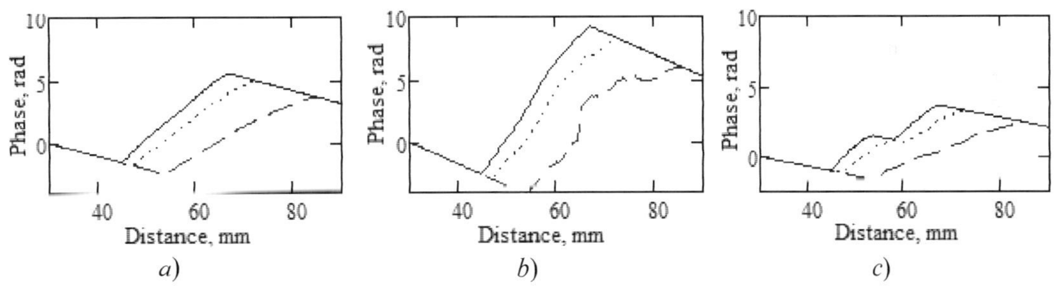

Figure 3. Phase spatial characteristics of analytical representations: *a*) RF 100 kHz, *b*) 200 kHz, c*)* square-top.

Figure 3 shows dependencies of phase characteristics of distance. Conditions are same as for the corresponding amplitude-time characteristics with same indexes. To determine the depths of the defect location t time aspect is replaced by spatial by multiplication of the time and sound velocity in specimen. As you can see the greatest gain slope is obtained with RF impulse of 200 kHz frequency.

The inflection of phase characteristic corresponds to distance at which excitation wave travels from transmitter to transducer, reflecting from the defect.

Under given geometrical conditions average distance is supposed to be equal to 47 mm. The result obtained from the model is 45 mm, which is close to the actual one.

4. Conclusion

The calculation shows that it is possible to use MET method to determine the location of the object knowing the exact position of CM and amplitude-time characteristics. Response signal phase characteristic allows evaluating depth of defect location as well.

Acknowledgments
This work was supported by The Ministry of Education and Science of the Russian Federation in part of the Science program (project 11.980.2017/4.6)

References

[1] Surzhikov V P, Khorsov N N, Khorsov P N 2012 *Russian Journal of Nondestructive Testing.* **48**(2)85–89
[2] Fursa T V, Surzhikov A P, Khorsov N N 2005 *Selected Proceedings of the 8th China-Russia Simposium on New Materials and Technologies* 138–140
[3] Surzhikov V P, Khorsov N N 2011 *Russian Journal of Nondestructive Testing.* **47**(10),687–690
[4] Surzhikov V P, Khorsov N N, Khorsov P N 2013 *Testing. Diagnostics* (9) 17–20
[5] Fursa T V, Khorsov N N, Romanov D B 1999 *Technical Physics. The Russian Journal of Applied Physics* **44** (10) 1175–1178
[6] Fursa T V, Dann D D 2011 *Technical Physics* **56**(8)1112–1117
[7] Surzhikov V P, Khorsov N N, Khorsov P N 2011 *Testing. Diagnostics* (11) 17–21
[8] Khorsov P N, Surzhikov A P, Surzhikov V P, Laas R A 2018 *Key Engineering Materials* **781** 155

IOP Conf. Series: Materials Science and Engineering **457** (2018) 012021 doi:10.1088/1757-899X/457/1/012021

Study of the technological process parameters basis on a mathematical model

I V Plotnikova[1], L A Redko[1], N V Chicherina[1], O N Efremova[1], S G Serebryakov[2]

[1]Associate Professor, Tomsk Polytechnic University, Tomsk, Russia
[2]Associate professor, Gubkin Russian State University of Oil and Gas, Moscow, Russia
E-mail: inna@tpu.ru

Abstract. The results of modeling of parameters of technological process are presented in this article. By means of data processing of an experiment the mathematical expressions connecting values of input and output parameters of process are received. On the basis of the regression analysis the system from three equations has been calculated. The check of the constructed model on suitability of application is executed. The received results will allow operating the process parameters so that to minimize the amount of discrepancies in the studied parameter that will lead to increase in production efficiency, decrease in level of marriage, decrease in grade of production, reduction of number of the controlling operations.

1. Introduction

Control and product quality are the main problems of modern production. The quality of modern products is considered as a complex indicator, which depends on the general scientific level of development, the quality of products, the perfection of technology and metrological support of production. Therefore, the problems of product quality should be solved at the earliest stages and stages of development of the technological process.

Any control system includes entrance, current and output control [1, 2]. The entrance control provides with big reliability of assessment of production quality shown on control; prevention of start in production or repair of production which isn't conforming to the established requirements. On the production which isn't conforming to requirements of standard documentation the act of defective production is made out. The current control represents the control of course of process by operators on each site of production which is carried out in the automated way [3–5]. The output control of finished production is carried out in production laboratory of the enterprise on compliance of quality indicators [6]. Unfortunately, the presence of automatic control does not guarantee 100% quality control of the assembled products. There are hidden defects in the joints (microcracks, air bubbles, unreliable contact of the hidden conclusions of the components BGA etc.), which are manifested in the operation of the nodes.

We will consider the line of impregnation of decorative paper with the melamine pitches as a result of which the high-quality melamine film applied for the lamination of the wood-shaving plates is made.

For the purpose of prevention of production of inappropriate products, we will exercise control of production process [7, 8]. For this purpose, we use mathematical model of process.

2. Research description

This article is the continuation of works [9, 10] in which stages of studying of this process are presented, namely:

- technological process is investigated;
- the correlation dependence between factors is constructed;
- the most significant factors are revealed;
- the plan of an experiment is developed.

The purpose of the following stage of this work was drawing up a system of the equations with use of the regression analysis on the basis of processing of the obtained data.

The object of a research is the technological process of the chipboard's production.

The subject of a research is the system of communications between adjustable parameters of technological process of the chipboard's production.

When studying aprioristic information, the analysis of data with use of the following tools of quality was carried out: control card; histogram; Pareto's analysis; Isikava's chart; control leaf; chart of dispersion of [11,12].

Baseline data is given in [10].

Considering the coefficients we will make the system of the equations:

$$\begin{cases} y = 687.88 - 2.75\cdot x_1 + 2.38\cdot w_2 + 6.13\cdot w_4 + 1.75\cdot x_1\cdot w_2 + 1\cdot x_1\cdot w_4 + 2.38\cdot w_2\cdot w_4 + \\ + 0.25\cdot x_1\cdot w_2\cdot w_4; \\ y = 686.70 - 3.8\cdot x_2 + 0.9\cdot w_5 + 1.8\cdot x_2\cdot w_5; \\ y = 686.83 - 1.42\cdot x_3 - 0.83\cdot w_1 + 4.42\cdot w_3 + 1.92\cdot x_3\cdot w_1 + 3.17\cdot x_3\cdot w_3 - 0.42\cdot w_1\cdot w_3 + 0.33\cdot x_3\cdot w_1\cdot w_3. \end{cases}$$

Some of the equation coefficients are insignificant, i.e. they have no significant effect on formations of a process response. In this regard, these coefficients which reflect the influence of a factor or the effect of factors' interaction are excluded from the equations. An inspection of the importance of coefficients was carried out by Styudent's criterion.

Therefore, it is possible to draw a conclusion that all factors have a significant effect on formation of density value. Coefficients only at one pair interaction of humidity of shaving and the temperature of pressing are insignificant and at threefold interaction of all factors – humidity of shaving, a ratio of layers, pressing temperature.

Insignificant coefficients are excluded from the equations. Thus, the final option of system of the equations has this form:

$$\begin{cases} y = 687.88 - 2,75\cdot x_1 + 2,38\cdot w_2 + 6,13\cdot w_4 + 1.75\cdot x_1\cdot w_2 + 2,38\cdot w_2\cdot w_4; \\ y = 686.70 - 3.8\cdot x_2 + 1.8\cdot x_1\cdot w_5; \\ y = 686.83 - 1.42\cdot x_3 + 4.42\cdot w_3 + 1.92\cdot x_3\cdot w_1 + 3.17\cdot x_3\cdot w_3. \end{cases}$$

After the calculation of model's coefficients, it is required to check the model for suitability. Such check is called check of model's adequacy.

In statistics a very convenient way of check of a hypothesis of model's adequacy is developed [13, 14].

The received value is compared to the tabular one. If the calculated value doesn't exceed the tabular one with the set confidential probability, the model can be considered as adequate. Therefore, the hypothesis of model's adequacy is accepted.

3. Interpretation of results

Each of the equations shows certain results of influence of entrance factors on plate density at a stage of formation of a wood-shaving carpet.

1. The considered factors: humidity of shaving (x_1), ratio of layers (w_2) and temperature of pressing (w_4):

$$y_1 = 687.88 - 2.75\cdot x_1 + 2.38\cdot w_2 + 6.13\cdot w_4.$$

In this equation all coefficients are significant, i.e. each of factors has significant effect on changes of a density indicator of chipboard. The coefficient at x_1 shows that at increase in humidity at one unit from value 2.25, density decreases by 2.75 kg/m^3 and vice versa, at reduction of humidity of shaving by each 1 % of 2.25%, plate density every time will increase by 2.75 kg/m^3.

At increase in value of a factor of a layer's ratio by 1% of 20.6% density of a plate will increase by 2 kg/m^3, at reduction of value of a factor of a ratio of layers by 1% of 20.6% density of a plate will decrease by 2 kg/m^3.

At increase in temperature of pressing on 1^{0}C from 197^{0}C, density will increase on 6 kg/m^3. At reduction of pressing temperature on 1^{0}C from 197^{0}C, density will go down on 6 kg/m^3.

2. The considered factors: tarring of an inside layer (x_2), pressure on the press (w_5):

$$y_1 = 686.70 - 3.8 \cdot x_2.$$

In this set the pressure on the press has no significant effect on values of density.

As for tarring of an inside layer, at increase in this parameter by 1% of 7.1% density decreases by 4 kg/m^3, and respectively on the contrary.

3. The considered factors: tarring of an external layer (x_3), speed of formation of a wood-shaving carpet (w_1), weight of a carpet (w_3):

$$y_1 = 686{,}83 - 1.42 \cdot x_3 + 4.42 \cdot w_3$$

In this case, the speed of carpet formation influences density indicator not significantly. At the same time, increase in tarring of an external layer by 1% from 11.6 leads to reduction of density of a plate by 1 kg/m^3. Increase in weight of a carpet by 1 kg from 131 kg corresponds to increase in density by 4 kg/m^3.

To determine the optimal conditions for the process, the gradient motion method is applied to each of the equations found. The Table 1 presents the optimal parameters of the production process.

Table 1. Optimal parameters of the production process

Input data \ Output data		Layer ratio, [%]	Pressing temperature, [^{0}C]	Density, [kg/m^3]
Chip Moisture, [%]	2.4	20.5	196	683
	2.2	20.7	197	693
	2.1	20.8	198	697

The Table 2 shows the parameters for optimizing the production process, obtained empirically.

Table 2. Parameters for optimizing the production process, obtained empirically

Input data \ Output data		Layer ratio, [%]	Pressing temperature, [^{0}C]	Density, [kg/m^3]
Chip Moisture, [%]	2.4	20.5	196	677
	2.2	20.7	196	697
	2.1	20.8	196	682

4. Conclusions

So, all technique of planning of an experiment which has been received in the practical way is described.

Before carrying out an experiment the aprioristic information by means of the analysis of process of chipboard production with use of statistical control methods has been studied. The financial losses connected with the production scrap have been as a result estimated. Stages of technological process of chipboard production at which there is the greatest number of defective plates are found. Types of defects which make 80% of all mass of defects are established. Types of defects which incur the greatest losses for the enterprise are found. Such defects are "short, friable plates". When studying these types of defects, the reasons of their formation in a production cycle are established.

On the basis of the regression analysis the system from three equations has been calculated. The check of the constructed model on suitability of application is executed.

Interpretation of results by the indication of interference of factors on a research object response in concrete numerical values is made.

This method has allowed establishing the influence of the chipboards of parameters regulated during the technological process of producti on on a quality indicator. Density of chipboard is taken for the quality indicator demanding the improvement and stabilization [15].

Due to the restrictions which are available in work the mathematical model has been developed for one stage of production – formation of a wood-shaving carpet. At the same time the studied parameters of technological process have been divided into three groups. Therefore, the constructed mathematical model represents system from three equations. With its help it is possible to improve the results received in work if to make an experiment on the production. At the same time one equation will be received that taking into account the interaction of all factors will fully reflect their influence on the chosen quality indicator.

References

[1] Redko V V *et al* 2016 *J. Phys.: Conf. Series* **671(1)** 012049 doi: 10.1088/1742-6596/671/1/012049
[2] Fedorov E M *and* Koba A 2016 *MATEC Web of Conferences* **79** 01082 doi: 10.1051/20167901082
[3] Nizhegorodov A I *et al* 2018 *IOP Conf. Ser.: Mater. Sci. Eng* **289** 012014 doi: 10.1088/1757-899X/289/1/012014
[4] Gavrilin A N *et al* 2016 MATEC Web of Conf. **7** 01078 doi: 10.1051/matecconf/20167901078
[5] Nizhegorodov A I *et al* 2018 *Bulletin of the Tomsk Polytechnic University, Geo Assets Engineering.* **329(5)** 67–75
[6] Evans J R 2005 *Total Quality: Management, Organization and Strategy* (South-Western College Pub: US)
[7] Obukhov S G *et al* 2017 *IOP Conf. Ser.: Mater. Sci. Eng* **189** 012008 doi:10.1088/1757-899X/189/1/012008
[8] Obukhov S G *et al* 2016 *MATEC Web of Conferences* **91** 01036 doi: 10.1051/matecconf/20179101036
[9] Vasendina E A *et al* 2015 *IOP Conf. Ser.: Mater. Sci. Eng.* **81(1)** 012070 doi: 10.1088/1757-899X/81/1/012070
[10] Plotnikova I V *et al* 2016 *IOP Conf. Ser.: Mater. Sci. Eng.* **132(1)** 012023 doi: 10.1088/1757-899X/132/1/012023
[11] Kumje H *et al* 1991 *Statistical method to improve quality* (Finansyistatistika: Moscow) (in Russian)
[12] Neave H R 1987 *Journal of the Royal Statistical Society Series D The Statistician* **36(5)** 561–570
[13] Siegel A 2011 Practical Business Statistics Academic Press 640
[14] Abishev N K *et al* 2016 *Rupkatha Journal on Interdisciplinary Studies in Humanities* **8(2)** 117–127 doi: 10.21659/rupkatha.v8n2.14

IOP Conf. Series: Materials Science and Engineering **457** (2018) 012021 doi:10.1088/1757-899X/457/1/012021

[15] Deming W E 1986 *Out of the Crisis: Massachusetts Institute of Technology* (Center for Advanced Engineering Study: Cambridge, USA)

IOP Conf. Series: Materials Science and Engineering **457** (2018) 012022 doi:10.1088/1757-899X/457/1/012022

The calculation of the magnetic circuit of salient-pole electric machines

K Zhumagulov[1], Ye Sarsikeyev[2], D Kapanova[3]

[1]Professor of Saken Seifullin Kazakh Agrotechnical University, Astana, Kazakhstan
[2]Senior lecturer of Saken Seifullin Kazakh Agrotechnical University, Astana, Kazakhstan
[3]Assistant professor of Saken Seifullin Kazakh Agrotechnical University, Astana, Kazakhstan
E-mail: sarsikeyev.ermek@yandex.ru

Abstract. This article deals with the development of theories of electromechanical energy converters with a lower level of external magnetic fields, a new method for calculating magnetic circuits and electromagnetic processes in the ferromagnetic space of electromechanical converters of electro technical complexes, oriented to the use of modern computer facilities, are new methods of differential investigation of the influence of various parameters of electric machines on the level of external magnetic fields. Beyond the engineering methods for calculating the magnetic circuit, it remains to determine the magnetic field in the grooves of the rotor, the relationship of this field with the field of the ferromagnetic part of the core. The article presents the results of research methods and calculations of magnetic circuits and modes of operation of explicitly salient pole machine. Developed a fairly simple method of calculating harmonic of the magnetic field in the groove of the rotor and the interpolar space, which provides the definition of the parameters of the magnetic field at the given points without calculation of this field in the whole region.

1. Introduction

In the conditions of intensive development of powerful electric receivers and their control systems based on electronic devices - microprocessors, computers, communication equipment, the interaction of these devices, their quality and reliability of teamwork is becoming increasingly relevant [1, 2]. One of the issues of ensuring their electromagnetic compatibility is the reduction of external magnetic fields (EMF) of electric machines (EM) at the design stage.

2. Main part

In this direction various schools and groups study a number of problems and problems: new technical solutions and developments aimed at reducing the level of external fields, improvement of electromagnetic compatibility of converters and control systems, technical and economic indicators [3, 4]. This article deals with the development of theories of electromechanical energy converters with a lower level of external magnetic fields, a new method for calculating magnetic circuits and electromagnetic processes in the ferromagnetic space of electromechanical converters of electro technical complexes, oriented to the use of modern computer facilities, are new methods of differential investigation of the influence of various parameters of electric machines on the level of external magnetic fields. Beyond the engineering methods for calculating the magnetic circuit, it remains to determine the magnetic field in the grooves of the rotor, the relationship of this field with the field of the ferromagnetic part of the core. More strictly, the problem of calculating the magnetic circuit and operating modes of the pole-pole machines can be solved with the use of a modern apparatus of field theory.

When developing the method for calculating the magnetic circuit and operating modes, the following assumptions are accepted [5]:

1. The magnetic field in the grooves of the rotor and the interpolar space is considered to be plane-parallel.
2. The magnetic field is assumed to be stationary, i.e. obeys the equations:

$$div\overline{B} = 0,$$
$$rot\,\overline{H} = \delta$$

3. Calculation of magnetomotive force on the yoke is made along the middle line.
4. The highest harmonics of magnetomotive force of armature winding and the phenomenon of hysteresis of core it is disregarded.

For exact calculation of external magnetic fields of electrical machines, it is necessary to specify and simplify as much as possible calculation of the electrical machines making magnetic field.

The article presents the results of research methods and calculations of magnetic circuits and modes of operation of explicitly salient pole machine.

The use of known methods of calculation based on theconcepts of magnetic field, parameters, coefficients ofreaction anchors at presenting magnetic circuit in the form ofa lumped element cannot ensure currently required accuracy.

The magnetic field in the air gap is approximated using discontinuous harmonic functions. Since the distribution curve of the magnetic induction along the circumference of the stator has the character of a discontinuous periodic function, it, like any periodic function can be represented as a Fourier series. Then, the calculation of intensities, magnetomotive force and magnetic induction, and the characteristics can be numerical harmonic iteration method (taking into account the peculiarities of calculation of individual sections of the magnetic circuit) with the approach of the magnetic induction via the harmonic components of the Fourier series. However, there is another approach to solving the problem. The idea is that a complicated distribution function of the normal component of the magnetic induction logically represented as a set of simple, each of which takes place in the width of the tooth (AB) or groove (BS), as shown in Figure 1. In turn, the distribution function of the normal component of the magnetic induction over the prong, and above the groove can be approximated by a discontinuous harmonic functions [6].

In this approach, the distribution curve of magnetic induction in the air gap, the width of the tooth of the rotor can be written as follows:

$$b_\delta = \sum_\rho B_{\delta\rho} \cdot \cos(\frac{\rho \cdot \pi}{N} \cdot x') \cdot F_p(1), \tag{1}$$

where = 0, 1, 2, 3 ... is the harmonic order x' - discrete coordinate values 0, 1, 2, 3, ..., N-1; N- is the number of equally spaced points on the width of one tooth, $F_p(1)=1$ the width of the tooth, $F(1)=0$ beyond.

The distribution of normal to the bore of the stator, the magnetic induction on the width of a groove of the rotor is written:

$$b_\delta = \sum_\rho B_{N\rho} \cdot \cos(\frac{\rho \cdot \pi}{N} \cdot x'') \cdot F_p(1), \tag{2}$$

where $B_{N\rho}$ – peak value of the cutoff harmonic on the width of a groove of the rotor; N - the number of equally spaced points across the width of the groove, equal to the number of points across the width of the prong.

As can be seen from Figures 1, 2 harmonic accurately describe the function $b_\delta(x)$.

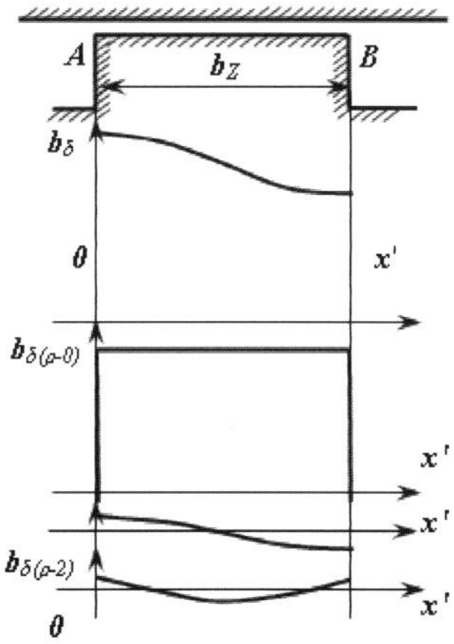

Figure 1. Approximation of the normal component of induction on the width of the prong explosive harmonic.

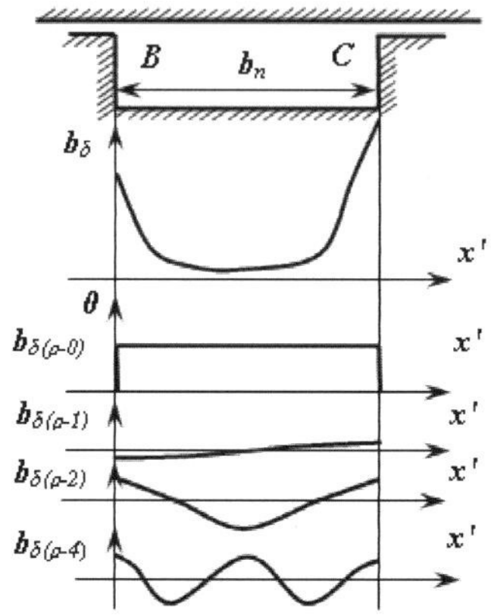

Figure 2. Approximation of the normal component in the width of the groove discontinuous harmonic.

The calculation of the cutoff harmonic of magnetomotive force of the magnetic circuit has a number of features. This is a special feature of the method of discontinuous harmonic. Breaking the zero-order harmonic ($\rho = 0$), which determines the constant component of the magnetic field, closes the circuit of the classical scheme of the magnetic circuit from pole to pole.

Magnetomotive force in the magnetic circuit for this component [7]

$$F_{n(p=0)} = F_{\delta(p=0)} + F_{z1(p=0)} + F_{z2(p=0)} + F'_{al(p=0)} + F''_{al(p=0)} + F_{a2(p=0)}, \quad (3)$$

where $F_{\delta(p=0)}$, $F_{z1(p-0)}$, $F_{z2(p=0)}$ – magnetomotive forces respectively in the air gap and the teeth of the stator and rotor; $F'_{al(p=0)}$, $F''_{al(p=0)}$ – magnetomotive forces in the yoke of the stator, respectively on a plot of the pole tip and winding; $F_{a2(p=0)}$ – magnetomotive force on the yoke of the rotor.

So bursting harmonic $\rho > 0$ closed on a completely different circuit than the harmonic $\rho = 0$, and the expression of the amplitude values of discontinuous harmonic magnetomotive force in a magnetic circuit for $\rho > 0$.

$$F_{n(p>0)} = F_{\delta(p>0)} + F_{z1(p>0)} + F_{z2(p>0)} + F_{al(p>0)} + F_{a2(p>0)}. \quad (4)$$

Magnetic fluxes from discontinuous harmonic at $\rho > 0$ closed at the pole tip and the body of the tooth of the rotor, as the yoke. While solving the problem of perforated layers each pole of the stator are replaced with the estimated homogeneous, the characteristic of magnetization of which in the radial direction coincides with the characteristics of the real magnetization layer, the perforated layer of the rotor is taken by real. Magnetization curve of steel is approximated by the specified points. Coefficient close m_{Z2} to one. Therefore, the steel teeth of the rotor are saturated. The penetration depth of the higher harmonic is small and it can be taken equal to half the pole division of the corresponding harmonic, or, in view of the smallness of tension in the rotor teeth, it is possible to neglect the magnetomotive force of the rotor to the higher harmonic.

For the constant component

$$f_{Zi(p=0)} = L_{Z2} \cdot H_{Zi(p=0)}, \quad (5)$$

where L_{Z2} – the height of the tooth of the rotor; $H_{Zi(p=0)} = f(B_{Zi(p=0)})$ – the magnetic field in i – the tine rotor, is determined by the magnetization curve of steel; $B_{Zi(p=0)} = m_{Z2} \cdot B_{\delta i(P=0)}$ - induction of the DC component in i – the tine rotor.

Magnetomotive force per air gap, the teeth of the stator and rotor

$$f_{\delta Zi} = f_{\delta i} + f_{Z1i} + f_{ri} \,. \tag{6}$$

Considering expression (6) as an even periodic function with half-period equal to the width of the tooth of the rotor, and decomposing it into Fourier series, the obtained amplitude values of discontinuous harmonic magnetomotive forces in the air gap and the teeth of the stator and rotor

$$F_{\delta Zi(\rho=0)} = \frac{1}{N}\sum_{x'=1}^{N} f_{\delta Zi} \,, \tag{7}$$

for harmonic $\rho > 0$

$$F_{\delta Zi\rho} = \frac{2}{N}\sum_{x'=1}^{N} f_{\delta Zi} \cdot \cos(\rho \cdot \frac{\pi}{N} x') \,. \tag{8}$$

3. Conclusions

The proposed approximation of the distribution curve of the normal component of the magnetic induction in the air gap by the method of discontinuous harmonic allows for enough high accuracy significantly reduce the range considered in the harmonic calculation. Piecewise-linear approximation of the normal component of the induction in the air gap discontinuous harmonic most closely corresponds to the idea of calculation of the different zones of the simplest - for each zone methods: field for non-ferromagnetic regions of the grooves of the rotor and numerical - harmonic - ferromagnetic. Developed a fairly simple method of calculating harmonic of the magnetic field in the groove of the rotor and the interpolar space, which provides the definition of the parameters of the magnetic field at the given points without calculation of this field in the whole region. The magnetic field in the slots of the rotor teeth and yokes, the output parameters which are a function of discontinuous harmonic magnetic induction over the rotor teeth, describes the independent variable systems of equations characterizing the modes of operation of salient pole machines [7].

Received program of the calculation of the magnetic circuit and operation modes of electric machines taking into account saturation in the presence of various types of eccentricity using a discontinuous harmonic functions. The program allows to calculate all quantities required for construction workers, energy and other characteristics, the curve of distribution of the normal component of the induction along the bore of the stator.

References

[1] Petrusev A, Rulevskiy V, Sarsikeyev Ye, Lyapunov D 2016 *Proceeding 2nd International Conference on Industrial Engineering, Applications and Manufacturing* **2016** 7911461 doi: 10.1109/ICIEAM.2016.7911461

[2] Obukhov S, Sarsikeyev Ye, Sholokhova I 2016 *Proceeding 2nd International Conference on Industrial Engineering, Applications and Manufacturing* **2016** 7911454 doi: 10.1109/ICIEAM.2016.7911454

[3] Burkin V, Kuznetsova N, Lopatin V 2009 *Technical Physics* **54** (5) 644–650 doi: 10.1134/S1063784209050065

[4] Voitenko N, Yudin A, Kuznetsova N 2015 *Journal of Physics: Conference Series* **652** 012011

[5] Zhumagulov K, Sarsikeyev Y, Kapanova D 2017 *MATEC Web of Conferences* **155** 01048

[6] Mendybayev S, Sarsikeyev Y, Kapanova D 2017 *MATEC Web of Conferences* **155** 01047

[7] Zhumagulov K 2006 *Calculation of the external magnetic field asynchronous machines with ferromagnetic screens* University's works of the KSTUn Karaganda **1** 80

IOP Conf. Series: Materials Science and Engineering **457** (2018) 012023 doi:10.1088/1757-899X/457/1/012023

Researches of air and fuel rate influence on oxygen level in emissions of new type medium power coal boiler

V Sergeyev [1], V V Yurchenko [2], S Zh Ayzhambayeva[1], G V Vavilova[3], M N Belik [2], S G Serebryakov[4]

[1]Associate Professor of Karaganda State Technical University, Karaganda, Kazakhstan
[2]Senior lector Karaganda State Technical University, Karaganda, Kazakhstan
[3]Associate Professor Division for Testing and Diagnostics of Tomsk Polytechnic University
[4]Associate professor, Gubkin Russian State University of Oil and Gas, Moscow, Russia
E-mail: jurchenkovv@mail.ru

Abstract. The article deals with sustained fire coal boilers requirements engineering from a creating energy saving and ecological compatibility point of view. The article gives experimental data obtained on boilers which was made by LLP «Карплаз», Karaganda, Republic of Kazakhstan.

1. Introduction

The research executed for high-power ecologically clean boilers requirements engineering and their certification paths. This research build upon usage of data obtained at LLP «Карплаз» which produce boilers of varying power for years. A coal has one major asset - it is domestic and as a result it is accessible and inexpensive.

Coal boiler is single acceptable alternative for heating. Coal has much the largest caloric power, is easy-to-use, and one coal load provides more long combustion period than wood or charcoal.

Conventional coal boilers require manual fuel load, they have fairly low efficiency and high level of carbon formation thereby flue needed to inspect and clean frequently. Air inflows bottom-up in natural way and fire bars are at the bottom of construction.

Introduction to prospective legislation of best available technology concept corresponds to accepted ecological legislation of EU countries approach to pollutant emission limitation from CHP. Republic of Kazakhstan ecological legislation harmonization with EU countries legislation is a development strategic direction of national legislation

If we focus on European legislation which enact employment of the best available technologies, then we can direct attention to gradual but significant reduction negative impact of heating enterprise pollutant emission on the atmosphere.

Regulation activity of emission reduction from organic fuel consumption installations performed in accordance with EU directives [1–3] by emission limitation of specific pollutant. These directives contain standards pollutants limit concentration into boiler plants flue gases emitted to atmosphere. For fulfill the regulatory requirements technologies (methods) included BAT references are used.

Therefore establishment of regulatory standards for specific pollutant emission from boiler plants is intended to phase reduce negative impact of CHP-plant to the atmosphere considering affordability and economic efficiency of BAT adoption [4], which provides these standards.

IOP Conf. Series: Materials Science and Engineering **457** (2018) 012023 doi:10.1088/1757-899X/457/1/012023

For example, according to Directive 2001/80/EC [2] sulfur dioxide (SO2) limiting concentration is taken to be 2000 mg/m3 under normal conditions for existing boiler plants burning solid fuel with heating rating up to 100 MW, linear decrease is from 2000 to 400 mg/m3 under normal conditions for boiler plants with heating rating 100-500 MW.. Both factors - affordability and economic efficiency are taken into account in this case.

It is clear that engineering solution for SO2 capturing have already been developed for low-power plants by that time. But with limited means to reduce sulfur dioxide emissions it is reasonable to equip by sulfur absorb plants large boiler plants first of all, where gross emissions reduction per unit of invested funds much greater.

Consequently, in accepting of Directive 2001/80/EC drastic reduction of CHP-plant negative impact was economically impractical with limited funds, because slight decline of gross emission at low power boiler plants demand for more costs.

The BAT concept in the EU as complex prevention and control of environmental pollution caused by anthropogenic activities provided by directives [5, 6] consider possible economic expenditures and environmental gains resulting from the implementation of BAT.

Fee collection for environmental pollution is kind of form of economic damage recompense that compensates impact of pollutants emission into the environment [7]. But at the present time in the Republic of Kazakhstan standard fee rates for emissions into the atmosphere of pollutants are such that in very rare cases emission fee will be greater than economic expenditures for introduction of new technologies and gas-cleaning equipment.

Despite the continuous technological development, which makes BAT precise definition difficult, it is still possible to choose the best technology among all available during relatively short period, which has been done in European BAT reference books. However, in order for determine the best technology in specific practical conditions criteria that depend on subjective decisions are necessary.

2. Research

Existing condition analysis of boiler plants burning coal, gas, fuel oil shows a large setting range of pollutant specific emission. These differences may result from main factors influence with boiler plant operation:

- boiler capacity;
- boiler operating life (and thus their technical condition);
- burn fuel structure;
- burn fuel quality;
- variety of implemented measures scope to suppress nitrogen oxide formation (or their absence);
- ARU technical condition and level of gas removal in them.

Specific emissions of harmful substances into the atmosphere from boiler plants must be control in order to audit of compliance with approved regulatory standards of specific emissions [8].

Specific emissions of i-th substance can be determined per unit input heat in firebox (g / MJ) or per ton of equivalent fuel (kg / ton of equivalent fuel) or evaluate as the concentration of this substance in 1 m^3 of fire gases under normal conditions and excess air coefficient $\alpha = 1.4$.

Specific emission standards for boiler plants are set for the following pollutants: nitrogen oxides (in terms of NO_2), sulfur dioxide, carbon monoxide, solid fuel ash.

Sulfur dioxide, nitrogen oxides and carbon oxides concentration is determined against the dry gases volume which is equivalent of measurements conditions of these substances by instrumental methods. Nitrogen oxides (NO_x) are calculated in terms of nitrogen dioxide (NO_2).

Specific mass pollutant emission into the atmosphere per unit of thermal energy input in firebox (g / MJ), mass pollutant emission per 1 ton of equivalent fuel is define as main regulatory limit.

Mass concentration of pollutant in flue gases released into the atmosphere in mg / m3 (at a temperature of 0 °C, a pressure of 101.3 kPa and $\alpha = 1.4$) is taken as a derived quantity.

Specific emission of i-th substance (n) in grams per megajoule, per unit heat input in firebox is determined by equation (1):

$$n = \frac{M_i}{BQ_i^r} \tag{1}$$

where M_i – emission value in grams per unit time (g/s, g/h, t/year); B – fuel rate for the same unit of time (kg/s, kg/h, kg/year, for gaseous fuel - m^3/s, m^3/h, m^3/year); Q_i^r – lower fuel calorific power, MJ / kg (for gaseous fuels – MJ / m^3).

Specific emission of i-th substance (m) in kilograms per ton of equivalent fuel or in grams per kilogram of equivalent fuel is determined by the equation:

$$m = n\ Qeq.fuel, \tag{2}$$

where $Qeq.fuel$ – equivalent fuel calorific power, equal to 29.33 MJ / kg.

The concentration of the *i*-th pollutant substance in fire gases (μ) in milligrams per cubic meter emitted into the atmosphere corresponding to specific index n is determined by formula (3):

$$\mu = n\frac{Q_i^r}{V_r} \tag{3}$$

where $V_г$ – fire gases volume under normal conditions in m^3 / kg (m^3/m^3 for gaseous fuels) and α=1.4.

At determining of sulfur oxides, nitrogen oxides and carbon monoxide concentration in milligrams per cubic meter dry gases volume is substituted into equation (3).

Specific emission regulatory standard at co-incineration of several fuels in boiler is defined as weighted mean value (4):

$$n_{cp} = \frac{\sum_{i=1}^{n} n_i \cdot B_i}{\sum_{i=1}^{n} B_i}, \tag{4}$$

where n_i – specific emission when operating on the i-th fuel, g / MJ; B_i – *i*-th fuel rate, g/s (g/year); $\sum_{i=1}^{n} B_i$ – total fuel rate per boiler, g/s (g/year).

Carbon monoxide maximum emission from the boiler is determined by the equation (5):

$$M_{CO} = \mu_{CO} \cdot \sum(V_{cri} \cdot B_i) \cdot 10^{-3}\ , g/s \tag{5}$$

where η_{CO} – measured carbon monoxide concentration in boiler exhaust gas at its maximum load when i-th fuel (or fuel mixture) burning, mg / m under normal conditions [9 –11].

The specific emissions values per unit of heat input in firebox per unit of burning equivalent fuel or per unit of emitted fire gases volume are specific emissions actual values corresponding to regulatory standard of maximum emission (MPE, g/s). Specific emissions actual values reflect attained level of boilers operation under projected burden at the winter maximum, take into account effect of previously implemented measures and represent pollutants specific emissions from these boilers that can be provided by current equipment operating and electricity and heat generating. These indicators are working tool for pollutants emissions control.

At switching to technological regulatory actions situation may arise when specific emissions actual values for many boiler plants exceed the established standards for boiler plants specific emissions.

Modern coal-fired boilers use technology that increases duration of heat carrier combustion which guarantees its full-fledged use. Figures 1–3 show a 300 kW boiler that has been tested at the coal mining enterprise of « Rapid » LLP. The boiler had a bunker for 1000 kg of coal, coal screw feeding (Figure 2). The air is fed by a fan at the screw feeder mouth. Coal feed rate and air flow rate are controlled by digital hand controls on the control panel.

Figure 1. Exterior of boiler with screw feed for 300 kW. Established at LLP «Rapid».

Figure 2. Boiler side-view - auger feeder for coal feed from a bunker.

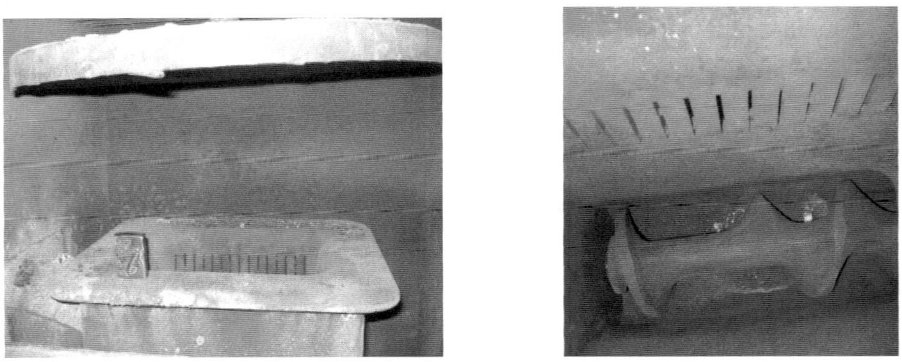

Figure 3. Firebox and screw exterior.

Following parameters were controlled:

- coal feed rate control;
- circulation rate;
- boiler jacket water temperature, on the feed and on the return;
- hot gases temperature at outlet from the boiler;
- outside air temperature;
- O_2 and CO_2 level in outgoing gases.

Testing purpose was the possibility of using this boilers type for reasons of energy saving and ecological compatibility. During the tests was established following:

IOP Conf. Series: Materials Science and Engineering **457** (2018) 012023 doi:10.1088/1757-899X/457/1/012023

- coal feed rate and circulation rate should be selected for each type of coal, otherwise the firebox is damped either from an excess of coal (the coal does not burn out poured) or from the excess air that blows the furnace;
- furnace is not suitable for long-term use, so the constant presence of a human being is required.

Portable gas analyzer Optima 7 was used as a control device for analysis of smoke and process gases, as well as for repugnant substance emissions control. It measures up to 7 gases simultaneously. This gas analyzer is ideally suited for boilers and turbines adjustment and control, and for ecological monitoring [12, 13].

Table 1. Controlled parameters.

Parameters	Measurement range
O_2	0...21 %
$CO (H_2)$	0 ... 4000 ppm short-term measurement up to 10000 ppm
CO low	0 ... 300 ppm
CO high	0 ... 4000 ppm
CO very high	0 ... 4,00% maximum up to 10%
NO	0 ... 1000 ppm
NO low	0 ... 300 ppm
NO_2	0 ... 200 ppm maximum up to 1000 ppm
SO_2	0 ... 2000 ppm maximum up to 5000 ppm
Tgas	0 ... 650 °C
Tenv.air	0... 100 °C
Draft / depression	-5 ... +35 hPa

3. Results and considerations

Table 2 shows measurements results in two modes - heating and temperature maintenance.

Table 2. Measurements results at heating and temperature maintenance modes

T_{pipe}, ^0C	40	60	80	100	120	143	147	133	122	110	94	115
T_{water}, ^0C	11	13	15	19	23	30	45	55	60	57	51	54
O_2, %	20.8	19	14.2	11.8	8.4	3.5	3.6	3.8	4.7	6	7.5	6.6
t, min	5	15	23	25	37	45	66	78	95	115	140	169
Note				Heating mode					Temperature maintenance mode			

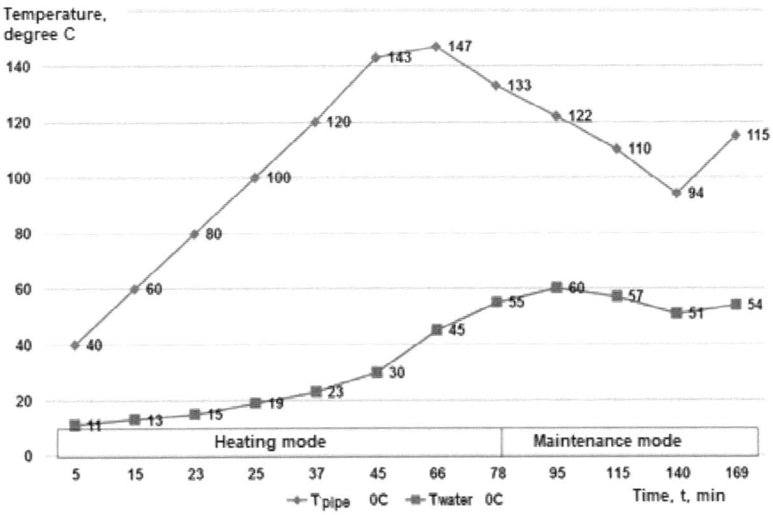

Figure 4. Schedule of boiler bringing into operation mode.

At performance and functional ability monitoring of complex dynamic objects represented by ensemble of a controlled object and an automatic control device, selection of controlled quantity and their normal values can be performed by analyzing differential equations describing the behavior of the system. In this analysis, of course, system purpose and criteria for evaluation its behavior are taken into account. Therefore, for automatically controlled objects, zones of normal values of quantities are often determined from requirement to maintain stability of standard conditions.

Controlled quantities selection can also be carried out in such a way as to satisfy specified control adequacy, time or cost criteria, etc.

Presence of measurement instrumentation imprecision leads to specific errors characterizing measurement quality.

There are errors of first kind which determine probability of assigning suitable control objects to unusable (not within the normal limits), and errors of second kind in the presence of which the unsuitable products are classified as suitable (within the normal limits).

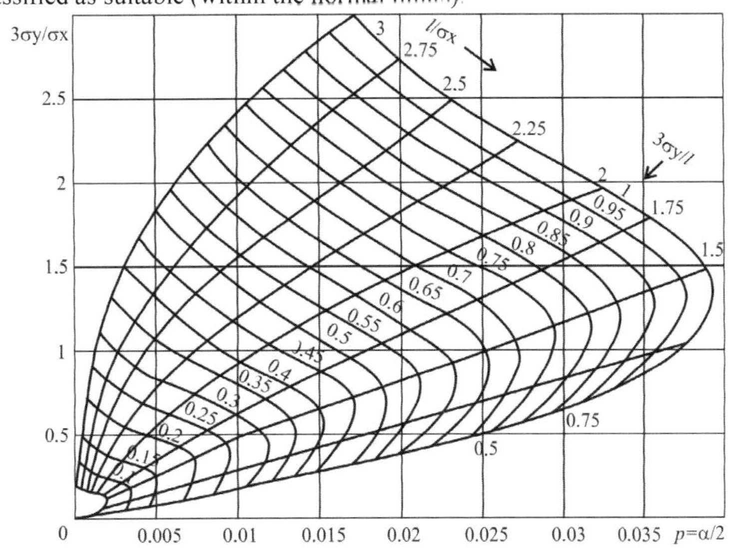

Figure 5. Errors of first kind probability, (vendor risk).

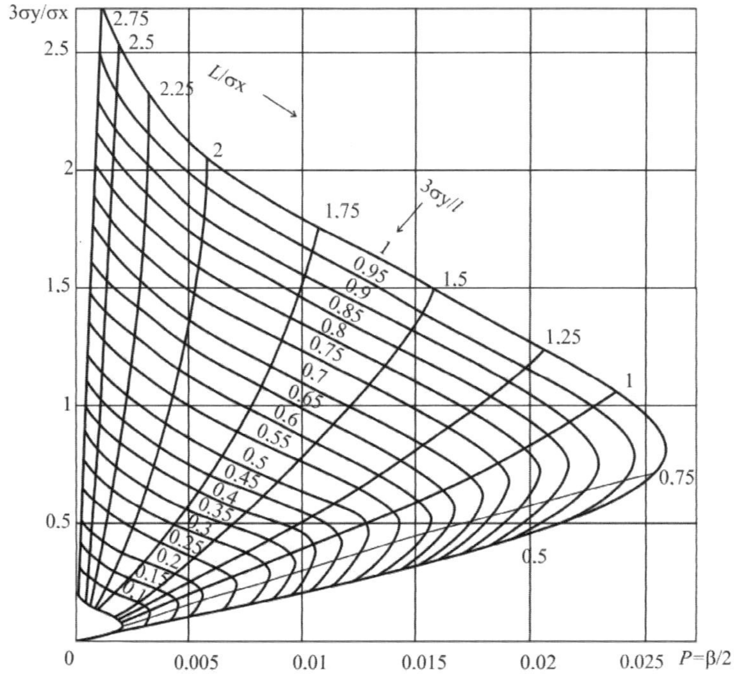

Figure 6. Probability of errors of second kind (buyer risk).

Ensured permitted allowance are used to drastic decrease in errors of second kind. This means that the tolerance zone c_u, c_l is narrowed to a size at which β becomes negligibly small. Obviously, this should increase errors of first kind. To ensure that the errors do not exceed the set value, it is necessary to ensure a determined precision of monitoring device [14,15].

Figure 6 shows a nomogram that allows to estimate necessary accuracy of the monitoring device if the tolerance zone is reduced by $3\sigma_y$, i.e., equal to $l-3\sigma_y$, errors of first kind α ratio of tolerance zone to mean square deviation of the controlled quantity are set.

Maximum error value of second kind does not exceed 10^{-4} for variation range in the accuracy characteristic of controlling device $3\sigma_y/l$ presented on nomogram. It is necessary to improve significantly standard accuracy to ensure that error of the first kind is no greater than at non-guaranteed error. Thus for $\alpha+\beta=0{,}04$, $l/\sigma x=2{,}5$ at non-guaranteed tolerance zone $3\sigma y/l\approx0.87$ (refer to nomogram at Figure 5), and for guaranteed tolerance zone [l—3σy] at $\alpha=0.04$, $\beta=0{,}0001$, $l/\sigma x=2.5$ device accuracy can be estimated by expression $3\sigma y/l\approx0.3$.

Therefore, control error check is necessary during the measurement procedure. It is desirable to take measures taking into account control errors of first and second kind. So as to exclude errors of the second kind, that is to not let devices that have metrological characteristics not corresponding to the normal ones specified in documentation be sent to consumer.

In our case, it is required to provide specified accuracy ratio of ordinary measuring instrument and required accuracy of control parameter. It is required to determine the errors of first and second kind.

Having set accuracy ratio, we define errors of first kind in the nomogram in Figure 5.

Accuracy ratio comes out of ratio from 1: 5 to 1: 3, we choose the average ratio of 1:4.

We will calculate tolerance zone value of normalized controlled parameter of 180 mgm.

Symmetric tolerance zone l of coordinate measuring Y of harmful impuritie detected content, mgm, not more than for converter with nominal value and normal conditions - temperature 0 ° C and pressure 101.3 kPa:

0°(when Y value is located in the range from 5 to 180 mgm) -

$$l=\pm(0.5+0.01Y)$$

from 40° up to 70° (when Y value is located in the range from 5 to 50 mgm) -

$$l_Z=\pm(2+0.03Z)$$

$$l_Y=\pm(2+0.03Y)$$

Under normal law of distribution of controlled quantity and specified probability 0,95 mean square deviation of the controlled parameter (6).

$$\sigma_X=l/h \quad (6)$$

Student's coefficient can be approximately set to 2, it follows that (7)

$$\sigma_X=1.15 \quad (7)$$

and at ratio of 1:4, selected mean square deviation of standard (8), (9)

$$\sigma_Y=\sigma_X/4 \quad (8)$$

$$\sigma_Y=1.15/4\approx 0,3 \quad (9)$$

To calculate errors of first and second kind probability it is necessary to calculate the relation (10)

$$\frac{3\sigma_Y}{l}=\frac{3*0.3}{2.3}=0.39 \quad (10)$$

And according to the nomograms 1 2 in Fig. 1 we find errors of first and second kind

$$\alpha=0.02$$

$$\beta=0.007.$$

Similarly, we find that errors of first and second kind at all points have the same values.

4. Conclusions

It is necessary to ensure firebox automatic ignition at reduction of gas temperature in the pipe or at O_2 increase.

There is a need to take certification work to confirm produced power and ecological cleanness of entrained emissions at high-power boilers manufacturers. There are two ways to control this:

- test operation at special stationary stand;
- tests on mobile calibration outfit.

References

[1] Directive 2001/80/EC on the limitation of emissions of certain pollutants into the air from large combustion plants 2001

[2] Directive 2010/75/EU on industrial emissions (integrated pollution prevention and control) 2010

[3] Directive 96/61/EC of the European Parliament and of the Council of 24 September 1996 concerning integrated pollution prevention and control 1996

[4] Nizhegorodov A I *et al* 2018 *IOP Conference Series: Materials Science and Engineering* **289** 012014 doi:10.1088/1757-899X/289/1/012014

[5] Directive 2008/1/EC of the European Parliament and of the Council of 15 January 2008 concerning integrated pollution prevention and control 2008

[6] Instructional and methodological guidance for environmental pollution payment collection, Order of The Ministry of Natural Resources and Environment of the Russian Federation dated January 26, 1993 (as amended on 12 July, 2011)

[7] Plotnikova I V, Chicherina N V, Bays S S, Bildanov R G, Stary O 2018 *IOP Conference Series: Materials Science and Engineering* **289** 012029 doi:10.1088/1757-899X/289/1/012029

[8] GOST (State Standard) R 50831-95: Boiler plants. Heat-mechanical equipment. General technical requirement, 1995

[9] RD 34.02.305-98. Methodology of Accounting Gross Polluting Substances Emissions from CHP

boiler plants 1998

[10] Plotnikova I V, Efremova O N, Tchaikovskaya O N 2017 Ponte **73**(4) 158–162 doi: 10.21506/j.ponte.2017.4.45

[11] RD 153-34.02.306-96. Rules on organization of air emissions control at thermal generation power plants and boilers. 1997

[12] Boiko E A *et al* 2009 *Boiler installations and steam generators: educational medium* (Tomsk Polytechnic University Publishing House: Tomsk)

[13] Vavilova G V, Gol'dshtein A E 2018 *Measurement Techniques* **61**(3) 278–283 doi: 10.1007/s11018-018-1421-6

[14] GOST (State Standard) R 50779.10-2000. Statistical methods. Probability and general statistical terms. Terms and definitions. 2000.

[15] Vavilova G V and Ryumkin A V 2018 *IOP Conference Series: Materials Science and Engineering* **289** 012017 doi: 10.1088/1757-899X/289/1/012017

IOP Conf. Series: Materials Science and Engineering **457** (2018) 012024 doi:10.1088/1757-899X/457/1/012024

Investigation of vibration impact on metrological characteristics of measuring devices in conditions of pump station №5 channel named after Satpayeva

S Zh Ayzhambayeva[1], D O Tyskenova[2], M V Vedyashkin[3]

[1]Associate professor, Measuring technics and instrument making, Karaganda State Technical University, Kazakhstan, Karaganda
[2]Assistant, Measuring technics and instrument making, Karaganda State Technical University, Kazakhstan, Karaganda
[3]Associate professor, Tomsk Polytechnic University, School of Non-Destructive Testing, Tomsk, Russia
E-mail: sauleaizh@mail.ru

Abstract. The article presents the results of the investigation of the effect of vibration on the metrological characteristics of measuring instruments, measurement methods and measurement techniques in real operating conditions. The topicality of the work is to ensure the reliability of the measurement results of the pump unit operation parameters in order to achieve the uninterrupted operation of the pump station objects. The investigation was carried out using the method 102-1, a vibration test was performed under the influence of sinusoidal vibration. The tests were carried out by smoothly changing the frequency in the specified range from lowest to highest. The conducted vibration studies made it possible to establish when the "normal" vibration becomes a "problem" or exceeds the permissible levels, the results obtained helped to determine the source and cause of the vibration, thereby ensuring effective prevention and eliminating the malfunctions of pumping units. In addition, it is established that when vibration occurs at several frequencies, it separates one frequency from the other so that it is possible to obtain each individual vibration characteristic. It is suggested at the pumping station to check the measuring instruments for vibration resistance, which allows to determine whether there are any additional errors of the instrument under the influence of vibrations during measurement, for example, as a result of tearing off the measuring tips of the CMM touch sensor from the measured surface. In the course of the study, we found that the use of vibration damping with the use of liquids for manometric equipment is quite effective in the conditions of operation of the pumping station. It is established that the most effective method is vibrodempfirovanie, which consists in reducing vibration by strengthening the design of friction processes that dissipate vibrational energy. This can be achieved by immersing the pressure gauge in the liquid, reducing the vibration amplitudes even in the resonance zone will depend on the damping ratio of the oscillations n..

1. Introduction

Regional State Enterprise "Kanysh Satpayev Canal" is intended for water supply of industrial regions and agriculture of Central Kazakhstan. The head water intake hub of the canal is located on the left-bank channel of the Irtysh River. White, above the city of Aksu. It ends at the pumping station No. 1 of the city of Karaganda, at 458 kilometers of the route. 272 kilometers of the channel pass through the territory of Pavlodar and 186 kilometers in the Karaganda region. Since 2001, water from the canal has been fed through the water pipeline to the Yesil river, then to the Vyacheslav reservoir for water supply to the capital of the republic - Astana. The idea of watering the arid Central-Kazakhstan region at the expense of the Irtysh River was expressed by many scientists long ago - back in 1930-1940.

The scheme of water supply of the city is divided into two zones because of the remoteness of the districts. The length of water supply networks in the city of Karaganda is - 1006 km. The design capacity of the water treatment facilities is 650 tons. The annual intake of water from the canal is 74 million cubic meters per year. The water from the channel to the scheme: channel - Pumping station № 1 - conduit - channel - Pumping station № 2 - penstock is fed to the water treatment plants, where the entire complex purification according to the standard. From wastewater treatment plants, drinking water is transported through 4 main water conduits to consumers. Such pumping stations – 6 pcs; booster pumping stations – 86 pcs.

2. Experimental

The main part of the pumping station is a pump unit [1, 2]. On. The pump unit consists of a centrifugal pump driven by an electric motor. The working member of the centrifugal pump - the impeller is installed in a ring-shaped chamber of variable cross-section [3, 4]. On the shaft of the impeller planted curved blades which during rotation of the pump entrain liquid fills the pump body, and the centrifugal force throws it by gradually expanding spiral chamber of the pump casing to the discharge nozzle. If in the case one wheel set, it is called a single-stage pump when multiple impellers placed in a common housing, the pump may be respectively two, three-stage, etc [5]. Due to the constant discharge of liquid in the rotating flow from the center of the pump wheel, a vacuum may be created in this zone, which is continuously replenished from the process pipeline by an external pressure at the pump intake [6, 7]

The technical operation of pumping units is the process of using them to fulfil the state plan for pumping volumes of water in order to guarantee the satisfaction of the needs of the economy while maintaining the technical parameters characterizing the technological process at the back level, ensuring maximum economy of operation, high efficiency, minimum cavitation damages of the flowing part and ensure the reliable operation of the channel as a whole. The technical characteristics of the OP pumps and the VDS electric motors of the Irtysh-Karaganda canal are given in Tables 1 and 2.

Table 1. Technical characteristics of the pump.

№	Name of parameters	Unit. amend.	Amount	
			OP10-185	OP11-185
1	Optimal pressure	m	21	17–18
2	Productivity	m^3/s	14–19	14–19
3	Rotational speed (slave)	rev / min	333	333
4	Direction of rotation	-	left	left
5	Blade turn range	-	0–9	0–8
6	Number of blades	pc.	6	4
7	Minimal penetration from the impeller	m	3.5	2.0
8	Coefficient of efficiency (max)	%	86	88
9	Angular rotation speed	rev / min	520	520
10	Diameter of pressure pipeline	m	2.6	2.6

Table 2. Technical characteristics of the electric motor – VDS 325/44-18.

№	Name of parameters	Unit. amend.	Amount
			OP10-185
1	Power	kW kVA	5000 5900
2	Voltage	V	6000
3	Stator current	A	566
4	Excitation voltage	V	to 150
5	Excitation current (rotor)	A	340
6	Power factor	-	0,9
7	Rotational speed	rev / min	333
8	Coefficient of efficiency	%	94.6
9	Causative agent	BBC 74/24-6	1
10	Exciter power	kW	76/5
11	Voltage	V	180/40
12	Weight VDS 325/44-18	H	50100

Instrumentation and automation facilities for the hydraulic engineering structures of the Irtish-Karaganda Canal, serviced by the KIP and A laboratories include:

1. Instruments for measuring water flow.
2. Devices for measuring the water level in the pool.
3. Instruments for measuring oil levels in engine oil sprouts VDS -325.
4. Devices for automatic monitoring and measurement of temperatures of electric motors, transformers;
5. Instruments for measuring pressure.
6. Vibration measuring instruments.

Oscillations (vibrations) in pumping units are considered primarily from the position of their influence on the durability of the operation of the elements of the design of any unit, the effect on the foundations or concrete overlaps where they are installed, and also on human effects [8–10]. The sources of vibration in pump units of water supply and discharge systems include: - pipelines; - pumps; - electric cars. Vibrations in pipelines are caused by wave processes occurring in the pumped liquid [11, 12]. During operation, the service life of radio electronic devices and systems installed at pumping stations is drastically reduced due to their operation under conditions of high random vibration. The increase in the periods of their working capacity and functioning is associated with carrying out highly effective vibration tests [13, 14].

Vibration has a different effect on the operation of devices and auxiliary devices. It increases the mechanical fatigue of parts and their wear, changes the frictional force in the rotating parts, increases the looseness of the movable joints, causes spontaneous unscrewing of bolts, screws and nuts. The most dangerous are the resonance and associated destruction of the individual elements and the whole structure, as well as the errors in the readings of the measuring instruments.

Low-frequency mechanical vibrations (2…10 Hz) can cause vibration of the whole device. The device should be isolated from the effects of these vibrations, otherwise high resolution can not be achieved. All these effects can be eliminated by careful design of the device, and therefore they are not very important when considering the limiting resolution. However, they are significant in the practical operation of the device [15 – 17].

IOP Conf. Series: Materials Science and Engineering **457** (2018) 012024 doi:10.1088/1757-899X/457/1/012024

3. Results and considerations

The study of vibration allows us to establish when the "normal" vibration becomes a "problem" or exceeds the permissible levels. These results can also help determine the source and cause of vibration, thereby providing effective prevention and troubleshooting. The vibration analyzer measures the amplitude, frequency and phase of vibration. In addition, when the vibration occurs at several frequencies, it separates one frequency from the other so that it is possible to obtain each individual vibration characteristic.

Instruments are tested in two ways for vibration and vibration resistance. Vibration test allows to identify drawbacks of assembly, soldering and other manufacturing defects. Therefore, it is advisable to check for vibration vibration all the devices at the factory. The vibration test allows you to determine whether there are any additional errors in the instrument due to vibrations during measurement, for example, as a result of tearing off the measuring tips of the CMM touch sensor from the measured surface.

Table 3. Comparative characteristics of measuring devices of pump station no 5.

№	Devices	Type		Class / non-pred.	Range / Limit.	Vibration resistance	Possible. Amplitude and frequency of vibrations.
1	Amperemetre	M325	Shielded	1.5	0…500	-	5.5
2	Voltmeter	M325	Shielded	1.5	0…250	-	5.5
3	Varmeter	C42303	Shielded	1.5	0…6	acceleration 15 m/s^2, frequency of 30 Hz.	5.5
4	Measuring t / n	VTOM-10	Stationary	0.5	57–6000	-	5.5
5	Current transformer	CTPCI-10	Stationary	0.5	2000/5	-	5.5
6	Manometer	STM4	Stationary	1.5	0…2.5	-	5.5
7	Flowmeter "Vzlyot MR"	UFM-510d	Shielded	-	-	-	5.5
8	Electricity meter	SA3U-E720	Stationary	1	-	-	5,5

It is advisable to check the instruments for vibration resistance only when developing and releasing new models. Electrodynamic vibration sets with a frequency range of 5-5000 Hz, a maximum displacement of 10 mm are suitable for testing instruments for vibration resistance and vibration resistance.

Verification of the vibration resistance of the device and evaluation of the quality of the installation of electrical equipment is carried out using a vibration shaker that generates vibrations at a frequency of 50 Hz and with an amplitude of vibration displacement (half the span) to 0.2 mm. All components of the device are fixed in the working position on the platform of the shaker. Within 10 minutes, tests are carried out on the modes specified in Technical Specifications [18–20]. The device is considered to have withstood the test, if, after the vibration load has ceased, the external inspection will not reveal any mechanical damage, weakening of the fastenings, a violation of the quality of the installation of electrical equipment. At the same time, the standardized characteristics of devices must meet the requirements of standards and technical conditions.

With the toughening of working conditions, the intensity and frequency of vibrations of aggregates, installations, etc. are increasing, just as pressure pulsations lead to an increased error in measuring pressure, wear of mechanical devices and, as a consequence, lead to a breakdown in the technological process, and therefore sharply increase requirements for vibration protection of sensitive devices for measuring pressure. As a protection against vibration, the following methods are used [14, 21]:

Vibration isolation: Reduction of vibration consists in reducing the transfer of vibrations from the oscillating device to the manometer, with the help of elastic pads, springs placed between them, etc. An example of vibroinsulation with Getzner gaskets is shown in Figure 1

Figure 1. Vibration isolation Getzner.

Vibration damping: is to increase the mass of the system and it is carried out by installing aggregates on a massive foundation [22, 23]. This method is effectively used when reducing the vibration of block equipment. In Figure 2 shows the vibration damping of a gas piston power plant.

Figure 2. Vibration damping.

Vibration damping: This method consists in reducing vibration by strengthening the design of friction processes that dissipate vibrational energy [24]. This can be achieved by immersing the pressure gauge in the liquid (Figure 3), reducing the vibration amplitudes even in the resonance zone will depend on the damping coefficient of the oscillations n.

IOP Conf. Series: Materials Science and Engineering **457** (2018) 012024 doi:10.1088/1757-899X/457/1/012024

Figure 3. Glycerin filled manometer.

4. Conclusions

Protection from vibration and shock is achieved through the convenient placement of devices, reducing the vibration of the source of vibration and the use of anti-vibration devices. Of course, it is best to install instruments in special rooms, away from sources of vibration and shocks. If this is not possible, it is desirable to mix the analyzers on massive shock-absorbing stands or independent foundations, or finally, near supporting structures (walls, columns), where the vibration is felt less. The balancing of moving parts and the multiplication of the oscillation frequency of a source above 10-20 Hz (devices are easier to protect from the effect of a higher frequency of oscillations) make it possible to reduce the influence of vibration of oscillation sources.

References

[1] Averyanov G C, Belkov V N, Buryanov Yu A, Korchagin A B, Komarov Yu P 2012 *Omsk Scientific Bulletin* 1(107) 43–46

[2] GOST 15467-79 – Upravlenie kachestvom produktsii, State Standard 15467-79 Product Quality Management (in Russian)

[3] Hummel K, Shirman S 2008 *Components and technologies* **3** 38-42

[4] Vavilova G V, Gol'dshtein A E 2018 *Measurement Techniques* **61** (3) 278–283 doi: 10.1007/s11018-018-1421-6

[5] Gavrilin A N, Moyzes B B, Cherkasov A I, Mel'nov K V, Zhang X 2016 MATEC Web of *Conf.* **79** 01078 doi: 10.1051/matecconf/20167901078

[6] State Standard 30630.1.2-99 Test methods for resistance to mechanical external influencing factors of machines, devices and other technical products (in Russian)

[7] Shirman A R, Solovyev A B 2013 *Practical vibrodiagnostics and monitoring of the state of mechanical equipment* (Moscow)

[8] Mitrofanov S S 2015 *Theoretical and physical basis of the device of optical instruments* (Saint Petersburg)

[9] Plotnikova I.V, Chicherina N V, Bays S S, Bildanov R G, Stary O *The selection criteria elements of X-ray optics system* 2018 *IOP Conference Series: Materials Science and Engineering,* **V. 289** doi:10.1088/1757-899X/289/1/012029

[10] Podolskyi V P 2010 *Volga Scientific Journal* **4**(26) 86–93 doi:10.18372/1990-5548.26.857

[11] Golubeva K V 2011 *Volga Scientific Journal* (4) 67–71

[12] Goldshtein A E Vavilova G V Mazikov S V 2016 *J. Phys.: Conf. Ser.* **671** 012062 doi: 10.1088/1742-6596/671/1/012062

[13] Pirogov S P Cherentsov D A Chuba A Yu 2015 *Oscillations of manometric tubular springs* (Tyumen State Oil and Gas University

[14] Ayzhambaeva S Zh Tuskenova D O 2016 *Proceedings of the International scientific conference Science integration, education and production –basis of the implementation of the Plan of the nation*

[15] Ayzhambaeva S Zh 2012 *Fundamentals of the theory of automatic control* (Publishing house Karaganda state university:Karaganda)

[16] Plotnikova I V Efremova O N Tchaikovskaya O N 2017 *Ponte* **73** (4) 158-162 doi: 10.21506/j.ponte.2017.4.45

[17] Salisbury C, Gillespie R B, Tan H, Barbagli F, Salisbury J K 2009 *Third Joint Eurohaptics Conference and Symposium on Haptic Interfaces for Virtual Environment and Teleoperator Systems* 115-120

[18] Clemens B M Kadis J Clemens D M Pollak E Clark P and Groves J R 2014 *Savart Journal* **1** (4) **1–9**

[19] Osawa Y Oguma Y 2013 *J Musculoskelet Neuronal Interact* **13(**4) 442-53

[20] Spreemann D, Manoli Y 2012 *Electromagnetic Vibration Energy Harvesting Devices. Springer Series in Advanced Microelectronics* **35** 107 doi: 10.1007/978-94-007-2944-5_5

[21] Cherentsov D A Pirogov S P Dorofeev S M 2014 *Bulletin of Tyumen State University* **7** 234-239

IOP Conf. Series: Materials Science and Engineering **457** (2018) 012025 doi:10.1088/1757-899X/457/1/012025

Objectivation of diagnostic indicator of defective insulating elements of strings by means of electric field intensity detection in inter-insulating intervals

L M Rybakov[1], N L Makarova[2], S V Lastochkin[3], A O Zakhvataeva[4], D V Rukavishnikova[4]

[1]Professor, Mari State University, Yoshkar-Ola, Russia
[2]Associate Professor, Mari State University, Yoshkar-Ola, Russia
[3]Aspirant, Mari State University, Yoshkar-Ola, Russia
[4]Master, Mari State University, Yoshkar-Ola, Russia
E-mail: diagnoz@marsu.ru

Abstract. The work describes existing methods of insulating elements diagnostic in strings. Non-traditional methods such as surface conductivity insulating monitoring methods, detection method of insulation current, heat monitoring method etc. were discussed. In the recommended method of heat monitoring with remote measurements temperature difference in insulator between correct and defective insulators is recommended 0.3-0.5°C that doesn't let to get true-to-fact data because of proximal meaning of uncertainty value to limiting resolution of thermal camera. Monitoring methods of insulation condition of overhead line contained in normative documents require correction and riders because of lack of information. Objectivation of diagnostic indicator for finding of defects in strips by means of electrical intensity detection in inter-insulating intervals was conducted. In the result of experimental research during modeling of the string containing 14 insults the presence of defect elements in different points along the strip with one, two, and three neutral insulators was imitated. 1467 experimental data were received from 16384 possible combinations. During the experiment electric field intensity in the air-gap between insulators and loss in voltage in insulators were measured off. Received electric field intensity curve shapes have equipotential character. On the base of the results calculation was made for each curve and numeric values of hyperbolic tangent at difference positions of neutral insulators in the string. On the received results graphs of change of *th(E/Δk)* at different value and position of saw-shaped neutral isolators were constructed. Comparison of received values of hyperbolic tangent at presence and absence of neutral insulators allows to find out the existence of defect elements in the string.

1. Introduction

Insulators condition on the transmission line and electricity distribution is a paramount important factor for an electric power system reliable functioning [1]. Damaged and inadequate to the standards for electrical characteristics defective isolators may lead to expensive shutdown electric system, serious injuries or fatal accidents [2].

For maintenance of overhead line, it is necessary to control of isolation condition for what different devices, methods and means can be used [3, 4]. Electromagnetic field, recording of optical emission and acoustic signals, temperature, and high-frequency electromagnetic radiation were used in applied methods as diagnostic indicators. Variety of used methods based on different physical features gives multivalued assessment and decrease in formativeness of the results. That is why, it is necessary to justify a diagnostic feature reacting to the change in electromagnetic fields when defective elements appear in strings at different numbers and position of the defect.

Alternating voltage distributes along insulators of the string irregularly and appearance of defective elements brings additional field distortions to inter-insulating interval may serve as diagnostic features that are not affected to external factors (temperature, illumination, acoustical signals).

2. Experimental

Currently, for overhead line control traditional and non-traditional methods are used.

2.1. The traditional methods. Traditional methods are considered in detail in volumes and norms of electrical equipment examination [5]. Thus, in the work considerable attention is paid to study of non - traditional methods of insulating monitoring.

2.1.1. Surface conductivity insulating monitoring method. This method is used for risk estimation of insulation contaminated layer of overhead line and choice of more effective lap preventive actions [6]. Knowing characteristics of contaminated layer, it is possible to calculate on some precision of flashover by means of numerical methods [7]. In the determination of contaminations characterized only by entrainment of manufacturing plants, it is possible to calculate contaminated layer thickness and conductivity on condition of thorough wetting, also moisture-spark voltage of insulation [8].

2.1.2. Detection method of insulation current. Substantial connection between moisture-spark characteristics and prestrike insulation currents appear due to equal insulator contamination [6]. However, insulation currents are highly correlated to insulator type, unequal contaminated layer and composition. Despite repeated attempts, insulating capability of insulators does not have correlation relationship with insulation currents [9].

2.1.3. Heat monitoring method. The method is based on distant determination of temperature of contact connection of cords and insulation on intensity and spectral content of infrared emission with use of thermal cameras and pyrometers.

When identifying line polymer insulators with rough defects for example with conducting extraneous bodies or wire in fiberglass rods the experience of thermal cameras appliance is rather illustrative [10].

Neutral insulators in strings may be as «colder» so «warmer» than correct insulators [11]. Pointed difference may be if an insulator has very low resistance. In this case heat power that excretes in it will be also very low and it will be colder than juxtaposed insulators. Conversely, if resistance of the insulator is enough high but lower than rejectable level of 3000 mega ohm it will be heated more than if it would be correct.

2.2. Methods of preventive insulation monitoring based on discharge process registry. Principal capability of diagnostic of overhead line condition method realizes on insulation with help of corona discharges intensiveness and surface partial discharge. In according to accepted standards of overhead line design field intensity on overhead conductors is restricted by $E = (0\ldots\ 0.9\ )E_{н,}$ where $E_{н,}$ is density of corona origin [12]. Defects, damages and structural breaking that appeared during operational process lead to uprising and reinforcement of intensity of surface partial discharge or corona discharges. Under development quantitative ways of corona discharges and surface partial discharge recording are based on special features of emission of different frequency range [13 –15].

Principally there are three possible detection method of corona discharges and surface partial discharge intensity: *acoustic, electromagnetic, optical.*

2.2.1. Acoustical way. Acoustic way consists of measuring acoustic noise generated by discharge processes mainly in the ultrasonic spectrum region. This way regarding overhead line insulation monitoring found limited *use because of deficient space resolving power capability, low interference protection and sensitivity*. Its scope is limited by distribution substations 6-35 kV [16].

2.2.2. Electromagnetic method. For purposes of insulation preventive monitoring some types of recording devices for electromagnetic emissions of corona discharges and surface partial discharge were worked out [17]. However, these inventions didn't become widespread because of low stability of measurement devices to corona discharges and different exterior interferences.

2.2.3. Optical method. Advantages of optical method of corona charges and surface partial discharge recording is high spatial-temporal ability measure of the resolution and sensitivity in comparison with other methods of detection [18–20]. Disadvantage of the method is the need for conduction of the insulator elements test during darkness hours, difficulty of surface partial discharge signals selection from corona discharge signals, also it is difficult to differ surface partial discharge moving along contaminated surface of correct insulator from motionless surface partial discharge positions connected with insulation faults.

3. Results of Researches

In high-voltage engineering laboratory of the Department of electric supply and technical diagnostics of Federal State Budgetary Educational Institution of Higher Education «Mari State University» experimental data on field-strength variation and insulated elements of strings consisted from 14 insulators with imitation of presence of defect elements in different points along the strip with one, two and three neutral insulators intension were taken off. 1467 experimental data were received from 16384 possible combinations.

Figures 1–3 show voltage distribution and electric field intensity along the string with one, two and three neutral insulators.

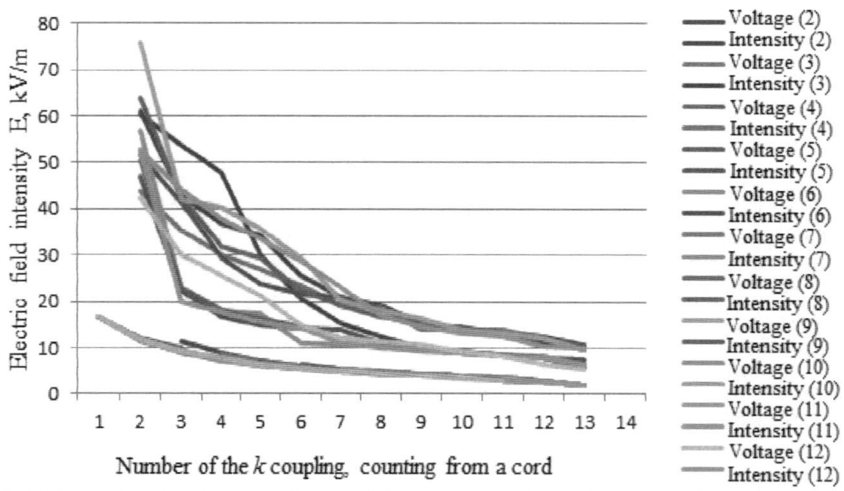

Figure 1. Voltage distribution and electric field intensity along the string with one neutral insulator in the string.

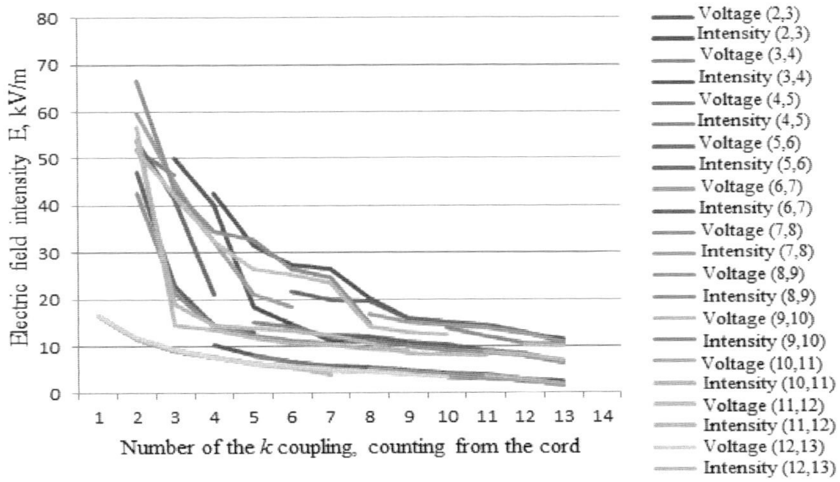

Figure 2. Electric-field voltage and intensity distribution along the string with 2 neutral insulators in the string.

The algorithm for finding defective elements based on the experimental data is as follows:

- since the shape of the electric-field intensity curves has an equipotential character (Figures 1– 3), it is possible to carry out a tangent to a curve from any point of the string of insulators (Figure 4);
- to calculate the hyperbolic tangent *th(E/Δk)*, from point 12, we draw tangents to the electric field intensity curves.

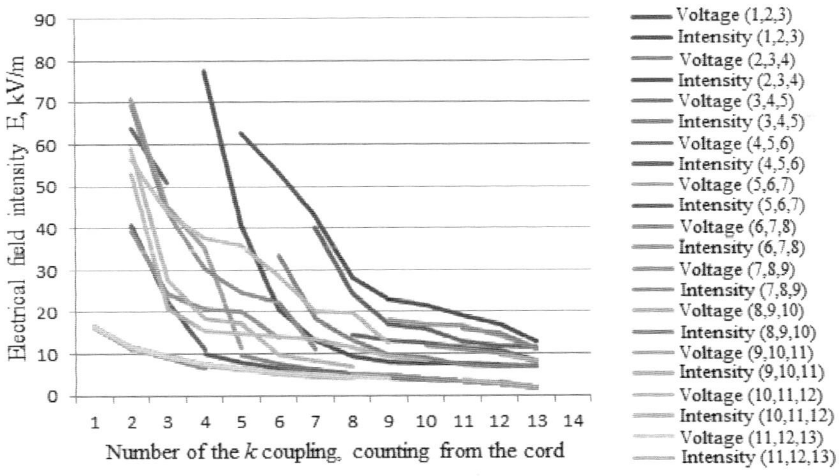

Figure 3. Electric-field voltage and intensity distribution along the string with 3 neutral insulators in the string.

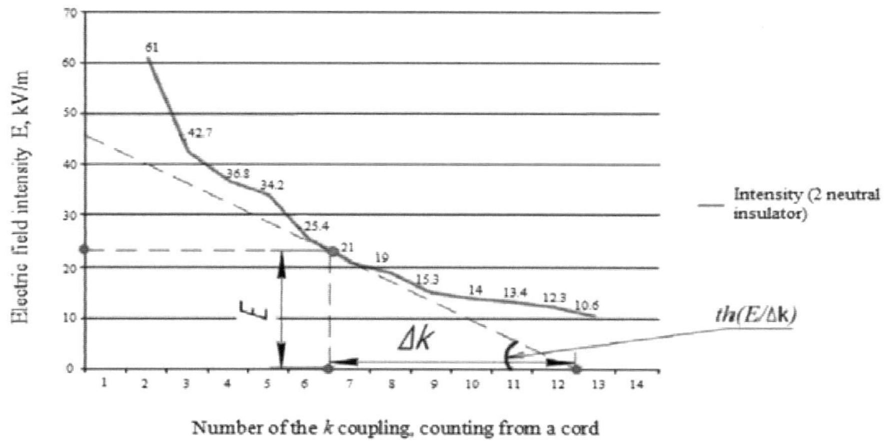

Figure 4. Determination of the hyperbolic tangent *th(E/Δk)*.

As an example, let us take the curve of the electric-field intensity distribution, where the second insulator, counting from the cord, will be the neutral one (Figure 4).

Figure 4 shows that *th(E/Δk) = th(21/(12-7))*= 0,999609.

Let us consider the cases when there are one, two, three zero insulators in a string (Figure 5, 6, 7). After the tangents to the curves of the electric field, obtain the numerical values of the hyperbolic tangent at different positions of the neutral insulators (Tables 1, 2, 3).

Table 1. th(E/Δk) value, with 1 neutral insulator in a string.

Number of insulator coupling, counting from the cord	*th(E/Δk)*	Number of insulator coupling, counting from the cord	*th(E/Δk)*
1	-	8	0.999295
2	0.999609	9	0.978026
3	0.995434	10	0.999831
4	0.99972	11	0.97627
5	0.969715	12	0.999593
6	0.997644	13	0.984921
7	0.950151	14	-

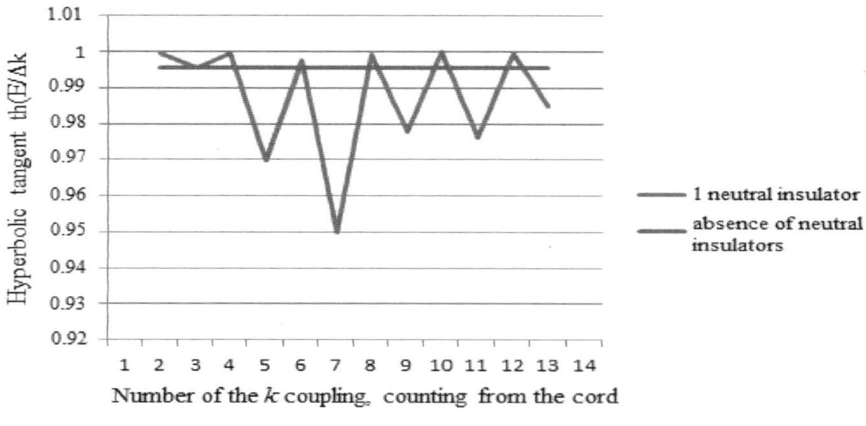

Figure 5. Change of *th(E/Δk)* at different positions of neutral insulators (1 neutral insulator).

Table 2. *th(E/Δk)* values with 2 neutral insulators in a string.

Number of insulator coupling, counting from the cord	*th(E/Δk)*	Number of insulator coupling, counting from the cord	*th(E/Δk)*
1	-	8	0.944155
2	0.977587	9	0.998838
3	0.99976	10	0.924946
4	0.973601	11	0.999393
5	0.999139	12	0.948085
6	0.949335	13	-
7	0.997898	14	-

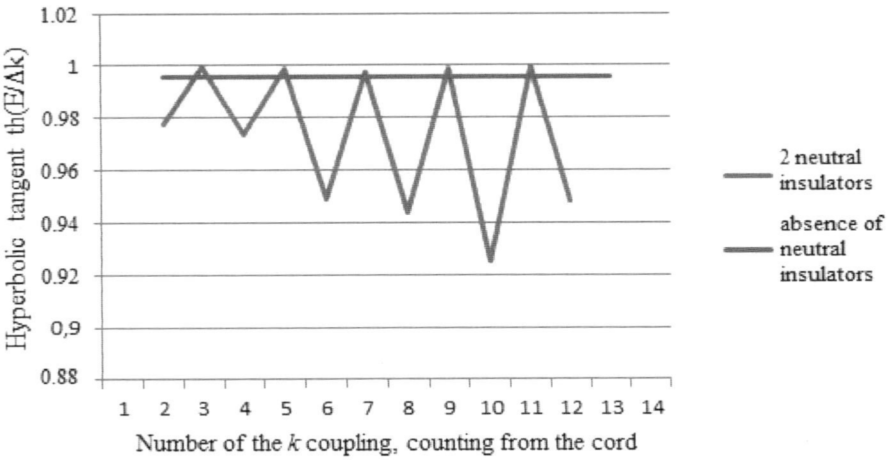

Figure 6. Change of *th(E/Δk)* at different positions of neutral insulators (2 neutral insulators).

Table 3. *th(E/Δk)* values with 3 neutral insulators in a string.

Number of insulator coupling, counting from the cord	*th(E/Δk)*	Number of insulator coupling, counting from the cord	*th(E/Δk)*
1	0.980096	8	0.999992
2	0.999999	9	0.919121
3	0.985508	10	0.99983
4	0.999978	11	0.959335
5	0.876972	12	-
6	0.997434	13	-
7	0.992126	14	-

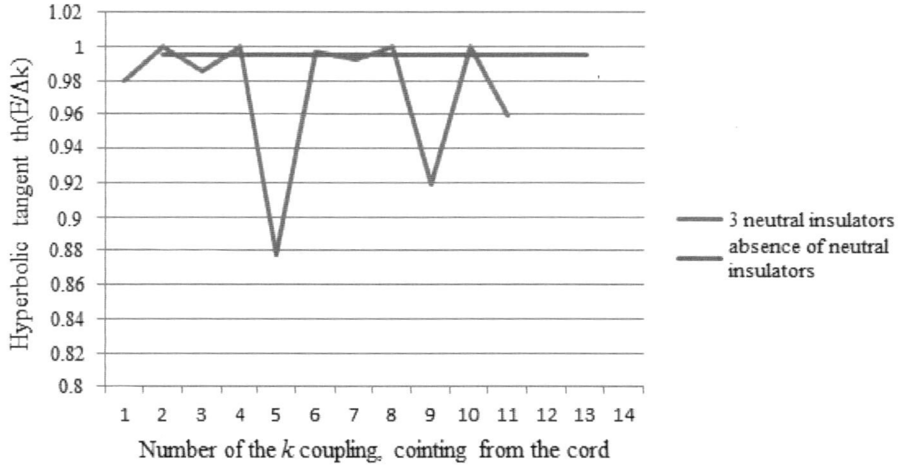

Figure 7. Change of *th(E/Δk)* at different positions of neutral insulators (3 neutral insulators).

As it can be seen from the graphs (Figure 5, 6, 7), the change of *th(E/Δk)* at different positions of the neutral insulators has the form of a sawtooth curve. Consequently, if we know *th(E/Δk)* without any neutral insulators, we can determine which element of the string has a defect. For example, Figure 5 shows that if the tangent value is higher than the straight *th(E/Δk)* without zero insulators, even insulators are defective, if the tangent value is lower than the straight *th(E/Δk)* without neutral insulators, odd insulators are defective. Thus, it is possible, by measuring the electric field intensity and changing the position of the measuring points (for example, to carry out measurements at different points of the string insulators) to obtain information about the presence or absence of defective elements.

The analysis of the search algorithm to find the defects in a string consisting of 14 insulators is carried out. According to the conducted analysis, it can be said that the search for defects in a string is possible by measuring the electric field intensity and calculating the hyperbolic tangent *th(E/Δk)*.

To develop the matter of finding defects in the overhead line strings, a method based on measuring the electric field intensity in the insulating intervals with the help of the PZ-50 device was chosen. The analysis of the experimental data showed that the search for defects in the string is possible by measuring the electric field intensity and, therefore, the calculation of the hyperbolic tangent *th(E/Δk)*.

4. Conclusion

1. The methods of monitoring the overhead line insulation condition, which are presented in normative documents, need updating and supplementing, as they have insufficient informative value.

2. Considering the proposed control methods, we recommend a remote method (thermal imaging method) that does not require shutdown of the equipment and does not affect its operation, as well as does not cause distortion in the current state of the controlled object. However, thermal imaging control method for remote measurement of the temperature difference on the insulator between serviceable and defective insulators is normalized 0.3–0.5 °C, which does not allow to obtain reliable values due to the close error value to the limit of the resolution of thermal imagers and the effects of climatic factors.

3. The substantiation of the diagnostic feature of defective insulating elements in strings by registration of electric field intensity in the inter-insulating intervals is carried out.

As a result of experimental studies on modeling defects in insulator string, equipotential curves were obtained, which allow determining the value of hyperbolic tangent at different number of neutral insulators in the string. According to the obtained values, a graph of *th(E/Δk)* changes at different positions of neutral insulators is constructed. The obtained graphs have a sawtooth shape. Comparison of the nature of the obtained hyperbolic tangent values in the absence and presence of neutral insulators helps to identify the presence of defective elements in the string.

References

[1] Cimador A *et al* 1994 *Proc of the 32th CIGRE Study Group Committee* **3** 04
[2] Devine C M *and* Farquhar J A 1985 *Electrical world* **199(12)** 59–61
[3] Lilien J L *and* Destine J 2012 *US Patent* 8184015
[4] Olsen R G *and* Edwards K S 2002 *IEEE transactions on power delivery* **17(4)** 1142–1152
[5] STO 34.01-23.1-001-2017 *Scope and norms of testing electric equipment*
[6] Merkhalev S D *and* Solomonik E A 1983 *The selection and operation of isolation in districts with contaminated atmosphere* (Energoatomizdat: Leningrad) (in Russian)
[7] Aksenov V A 1981 *Development of calculation methods and research of the prestrike characteristics and voltage overlap of polluted insulators* PhD Thesis (Sibir Energy Research Institute: Novosibirsk) (in Russian)
[8] Rutskiy V M 1994 *Development of a method for the selection of insulation levels in the area of the projected industrial enterprises* PhD Thesis (Sibir Energy Research Institute: Novosibirsk) (in Russian)
[9] Bazutkin V V *et al* 1986 *High voltage* (Energoatomizdat: Moscow) (in Russian)
[10] CIGRE WG 22.03 1996 Review of "In service diagnostic testing" of composite insulators *ELECTRA* **169** 105–119
[11] Taloverya V L *and* Chernov V F 1982 *Proc the all-Union Scientific and Technical Seminar* 38–42 (in Russian)
[12] Aleksandrov G.N. and Peterson L.L (eds) 1983 *Design of power transmission lines of ultrahigh voltage* (Energoatomizdat: Moscow) (in Russian)
[13] Tourrel D C *and* Ishivari M 2000 *Proc of the 38th CIGRE Study Group Committee* **1** 04
[14] Privalov E E 2015 *Diagnostics and thermal imaging control of electric power equipment: tutorial* (Direct-Media: Moscow-Berlin)
[15] Kalicki T *et al* 1995 *Proc of the 9th ISH* 5066
[16] Karkkainen S 1974 *Physical mechanisms of partial discharges* (VTT Technical Research Centre: Finland)
[17] Svi P M 1961 *Electric stations* **12** 34–36 (in Russian)
[18] Zhultz H *and* Zimmer H 1980 *Elektrizittswirtschift* **19** 704–708
[19] Sibiryakov V G 1985 *Development of methods and equipment for remote optical control of high-voltage isolation of overhead line and SWYD* PhD Thesis (Sibir Energy Research Institute: Novosibirsk)
[20] Benjamin J *et al* 1999 *IEEE Trans Dielectr Electr Insul* **6** 896 –899

AUTHOR INDEX